REGAINING SECURITY - A GUIDE TO T~~HE~~
DISPOSING OF PLUTONIUM AND HIGH~~LY~~
URANIUM

For Kierstan, Lanaya, Shelby and Rachel, my grandchildren whose future depends on a successful resolution of this problem.

Regaining Security - A Guide to the Costs of Disposing of Plutonium and Highly Enriched Uranium

WILLIAM J. WEIDA
Department of Economics
The Colorado College

ECCAR
Economists Allied for Arms Reduction

Routledge
Taylor & Francis Group
LONDON AND NEW YORK

First published 1997 by Ashgate Publishing

Reissued 2018 by Routledge
2 Park Square, Milton Park, Abingdon, Oxon, OX14 4RN
711 Third Avenue, New York, NY 10017, USA

Routledge is an imprint of the Taylor & Francis Group, an informa business

Publisher's Note
The publisher has gone to great lengths to ensure the quality of this reprint but points out that some imperfections in the original copies may be apparent.

Disclaimer
The publisher has made every effort to trace copyright holders and welcomes correspondence from those they have been unable to contact.

A Library of Congress record exists under LC control number: 97072265

ISBN 13: 978-0-367-00015-8 (hbk)
ISBN 13: 978-0-367-00016-5 (pbk)
ISBN 13: 978-0-429-44499-9 (ebk)

Contents

Figures and tables   x
Acknowledgments   xiii
Preface and executive summary   xiv

1 Economic issues and the disposition of weapon-grade plutonium and highly enriched uranium   1

Introduction   1
Disposition and the allocation of national resources   1
Non-proliferation   3
The goals of disposition   5
The negative externalities of wastes generated by disposition   7
The time frame for the disposition decision   8

2 The economic value of plutonium and highly enriched uranium   11

Economic value   11
Plutonium as an asset or liability   12
Disposition and economic value   13
Specific cost considerations for reprocessing during disposition   15
The value of recycled nuclear reactor fuel without reprocessing costs   17
The value of recycled nuclear reactor fuel with reprocessing costs   17
Russian perceptions of the value of plutonium   19
Additional costs associated with plutonium use and storage   20
The energy value environment   21

3 Disposition quantities of surplus materials   24

Introduction   24

Total quantities of plutonium   25
   US supplies of plutonium   25
   Russian supplies of plutonium   28
Status of materials in other countries   29
   Japanese supply of surplus plutonium   29
Total quantities of HEU   30

4 The economic nature of the US commercial nuclear industry   33
and the use of downblended HEU as reactor fuel

Classical economics and the commercial nuclear industry   33
The price of uranium   35
Burning plutonium in commercial reactors   36
Downblending highly enriched uranium for disposition as reactor fuel   36
   The mechanics of downblending   37
   The economics of downblending   37
Downblending options   41
Burning downblended HEU in light water reactors   44
Summary - disposition of HEU as reactor fuel   44
Decontamination and decommissioning costs   44

5 The impact of reprocessing on disposition   48

Introduction   48
Reprocessing   49
European and Japanese experience with reprocessing costs   53
   France   54
   UK   54
   Japan   55
Reprocessing plant capital costs   55
Interest charged during construction of reprocessing plants   57
Operating costs   59
Profits from reprocessing   60
Conclusion: economics of reprocessing - the European experience   60
US estimates of reprocessing costs   63
   Capital costs   63
   Operating costs   64
   Decommissioning costs   64
   Interest costs   64
Summary: a comparison of published reprocessing prices   68
Waste generation and US defense nuclear reprocessing   69
Alternative reprocessing technologies for separating plutonium only -
proliferation concerns   71

6 Disposition through MOX burning: fabrication and operations issues

6 Disposition through MOX burning: fabrication and operations 74
issues

Introduction   74
The uranium market   75
MOX use and subsidies   76
    Utility interest in MOX   76
    The need to subsidize MOX use   78
    Specific subsidy calculations   81
    Subsidies to investment in facilities   82
    Subsidies to operations of facilities   82
    Subsidies to capital costs   83
    Total life-cycle subsidies required by a MOX user - the sum of
    all subsidies in all categories   83
    Offsets to subsidy costs - fuel displacement credits   83
    Total subsidies required by MOX users with credit for fuel displaced   84
    Additional MOX subsidy issues   84
    Other estimates of MOX facility and MOX use costs   85
Capacity and locations of MOX facilities   86
Fabrication of MOX from surplus weapon plutonium - special problems   87
Civilian MOX fabrication   88
    France   88
    Germany   88
    Belgium   89
    Canada   89
    Great Britain   89
    Japan   89
    Russia   90
    United States   90
MOX utilization   90
    France   90
    Germany   91
    Switzerland   91
    Belgium   91
    Japan   91
Plutonium consumption rates in MOX   92

7 Burning plutonium in reactors: implications for disposition   97

Introduction   97
Developmental and alternative technologies for burning plutonium -
breeder reactors   99
    Advanced liquid metal reactors   99
Estimated costs of various disposition options   101

Specific experience with plutonium fuel use   104
   France   104
   Germany   105
   Belgium   106
   United Kingdom   106
   Canada   106
   Japan   106
   Russia   107
Conclusion: specific cost additions arising from the use of plutonium
in commercial reactors   108

8 A comparison of proposals for disposition of plutonium from warheads   112

Introduction   112
Disposition forms   113
   Highly enriched uranium (HEU)   113
   Plutonium   114
Disposition options that appear to be non-viable for technical or
economic reasons   116
   Underground fissioning by nuclear explosion   116
   Burning plutonium in light water reactors (LWRs)   116
   Burning plutonium in fast reactors   117
   Burning plutonium in unconventional matrices   117
   Launching plutonium into the sun   118
   Ocean-based disposition   118
     Sub-seabed disposal   118
     Dilution   119
     Tectonic Plate Burial   119
   Mix and melt disposition   119
   Transmutation   120
Disposition options that appear to be viable on both economic and
technical grounds   122
   Direct disposal in surface storage   123
   Mined geologic disposal   124
   Deep borehole disposal   126
   Vitrification   126
     Fabrication costs of vitrification   129
     Storage costs for vitrified material   131
Summary - a comparison of disposition costs   131

9 Conclusion: the major cost drivers in disposition   137

Introduction   137

HEU disposition   137
Plutonium disposition   138

Appendix 1   142

Cases where economic conversion serves as a rationale for selecting
specific methods of plutonium disposition   142
    Introduction   142
    Conversion as a general principle   143
    The Isaiah project   144
    The triple play reactor   145
    Conclusion   147

Glossary   149
Bibliography   155

Figures and tables

Table 3.1   World-wide totals of separated plutonium   24

Table 3.2   US plutonium inventories and locations   26

Figure 3.1   Dismantlement rates for nuclear warheads at the Pantex plant, Amarillo, Texas   27

Table 3.3   Pantex dismantlement as a percent of total workload   28

Table 3.4   Japan's separated plutonium inventory - end of 1994   30

Table 3.5   The US inventory of HEU - locations and amounts   31

Table 4.1   Blending requirements and yields - HEU   37

Table 4. 2   Cost summary of HEU disposition alternatives   39

Table 4.3   The value of 500 MT of downblended HEU   40

Table 4.4   Energy available from 1000 tons of HEU used as reactor fuel in 2005   43

Table 5.1   Euphemisms for reprocessing   51

Table 5.2   World oxide fuel reprocessing capacity   54

Table 5.3   Capital costs of the components of reprocessing plants   56

Table 5.4   Capital costs by reprocessing plant - a summary of estimates   57

Table 5.5   Total interest costs during reprocessing plant construction   58

Table 5.6   Reprocessing plant operating costs (based on a 900 MTHM/yr   59
annual throughput plant)

Table 5.7   General comparison of reactor fuel costs   61

Table 5.8   VVER-1000 fresh LEU fuel costs ($/kgHM)   61

Table 5.9   VVER-1000 fresh MOX costs ($/kgHM)   62

Table 5.10   German reprocessing costs compared with final storage
costs (1996 dollars per metric ton of spent fuel)   63

Table 5.11   Unit costs for reprocessing plants   66

Table 5.12   Budgeted amounts for separation and transmutation
technologies   67

Table 5.13   Summary costs: reprocessing price per kg heavy metal   67

Table 5.14   Spent fuel inventories at various storage sites   70

Table 5.15   Waste generation under various alternatives at the
Savannah River site - in cubic meters   71

Table 6.1   Utility and private company interest in MOX burning
and/or tritium production   77

Table 6.2   1996 DOE-generated costs for MOX use - millions of 1996 dollars   80

Table 6.3   Recent estimates of MOX facility and MOX burner costs - millions   86
of 1996 dollars

Table 7.1   Plutonium consumption rates for Russian reactors   98

Table 7.2   Generic capital, operations and maintenance costs for
various reactor types - 1996 dollars   102

Table 7.3   Approximate costs to dispose of 50 metric tons of
plutonium - 1996 dollars   103

Table 8.1   Potential transmutation technologies costs of
transmutation and other non-burning or technical fixes   121

Figure 8.1   The costs of canister fabrication at various production rates   130

Table 8.2   Summary comparison - cost estimates of disposition options -   132
            billions of 1996 dollars

Acknowledgments

This book represents the first publication in a Peace Economics Studies Series put out by Economists Allied for Arms Reductions (ECAAR). I would like to thank the members and board of ECAAR for their support and encouragement, and I would particularly like to acknowledge the help of Professor Walter Isard and the kindness shown me by Dr. Robert Schwartz, Professor Lawrence Klein, Professor Kenneth Arrow, and Professor James Tobin during the period of research that culminated in this book. Special thanks also goes to the W. Alton Joncs Foundation, the Unitarian Universalist Veatch Foundation, the Threshold Foundation, The Rubin Foundation, the Ploughshares Foundation, the Deer Creek Foundation, the Town Creek Foundation, and the Compton Foundation, all of whom have supported my research in the disposition of nuclear materials. Finally, it is especially important to acknowledge both Ms. Alice Slater, whose organizing and managerial abilities made this project possible, and the hundreds of citizen activists at sites around the United States whose efforts to control the hazardous substances discussed in this book have been singularly important in averting the terrible problems these materials could have caused.

Preface and executive summary

This book shows that costs of disposing of nuclear materials are most likely to be increased by subsidies to disposition programs, by increased volumes of waste from reprocessing, by increased handling of nuclear materials, and by technologically unproved methods of disposition. For economic reasons, plutonium is not a viable fuel for commercial reactors, and highly enriched uranium downblending faces the same unfavorable future as the nuclear power industry. Transmutation, which has high technological uncertainty and uses reprocessing is also not viable. This leaves only vitrification and a few surface or geological disposal methods as economically viable methods for disposition - the choice between them can be based on time, security and environmental concerns.

Because all aspects of decisions concerning disposition of nuclear materials are technically difficult, every effort has been made to explain the economics of disposition in a manner that is understandable to the informed lay person. To assist in making economic comparisons between disposition techniques, all costs in this book have been converted to 1996 dollars. In addition, the executive summary section that follows will act both as a general guide to the content of the chapters and as a preview of the major conclusions and findings of the book.

A complete glossary is included in the back of the book to assist the reader with the many specialized terms and acronyms used to describe nuclear issues, and a collection of costs for commonly discussed disposition options and factors likely to be the major cost drivers in disposition are both presented in the final chapters.

1. Economic issues and the disposition of weapon-grade plutonium and highly enriched uranium

Disposition involves dismantling weapons and warheads, intermediate disposition of nuclear materials, and long-term disposition of all surplus weapon-usable fissile materials.

There is currently no way to economically use plutonium as reactor fuel.

Plutonium of virtually any isotopic composition can be used to make nuclear weapons.

Over the next 10 years about 200 tons of plutonium and 1,000 tons of highly enriched uranium will be recovered from dismantled weapons.

There will be enough reactor-grade plutonium by 2010 to make about 71,000 primitive bombs

2. The economic value of plutonium and highly enriched uranium

Highly enriched uranium and plutonium are not uniquely valuable fuels for power reactors.

A fuel fabricator will pay for low enriched uranium instead of accepting free weapon-grade plutonium.

There is no viable market in weapon-grade plutonium - prices for plutonium and the amount of highly enriched uranium allowed on the market are set to achieve political, not economic ends.

Reprocessing is desirable only if there is a shortage of natural uranium at reasonable prices. Such a shortage is unlikely to ever occur.

Low enriched uranium costs about half the estimated cost of new plutonium (mixed oxide) fuel. Reactor burning adds about $1.7 billion to the costs to dispose of 200 tons of plutonium.

Plutonium becomes economically viable as reactor fuel only when uranium prices sharply increase. However, such a price rise would make all nuclear power plants unable to compete with competing sources of energy.

3. Disposition quantities of surplus materials

The 'pit' or explosive center of a nuclear warhead contains about 3-4 kg of plutonium. The uranium 'secondary' has about 15 kg of highly enriched uranium.

There are about 100 metric tons of weapon grade plutonium in both Russia and the US, with lesser amounts in Japan, Great Britain, and France.

There are about 2300 metric tons of highly enriched uranium worldwide, almost all of it in the former Soviet Union and the US.

By the year 2000, the total amount of separated plutonium in the civilian sector is expected to surpass the total amount in weapons arsenals.

The DOE's dismantlement goal of 2000 warheads per year has never been met.

4. The economic nature of the US commercial nuclear industry and the disposition of downblended highly enriched uranium as reactor fuel

From 1950 to 1990, 20% of the $595 billion spent to develop and obtain nuclear power came from the US government. Most federal funds were subsidies to research and development that resulted in a multitude of new reactor designs.

Failure to use a common reactor design has complicated regulatory problems and created economic inefficiencies across the commercial nuclear industry.

Plutonium use in commercial reactors will not be economically feasible for 50 years until the price of uranium-bearing yellowcake reaches $110/LB.

Fast breeder reactors will not be profitable until yellowcake prices reach $240/LB.

Highly enriched uranium can be downblended easily, but disposing of significant amounts of highly enriched uranium by downblending will lower demand for other US-fabricated low enriched uranium. Releasing the highly enriched uranium from dismantled nuclear weapons over a 10 year period could significantly disrupt the uranium market and result in facility closings.

One kilogram of highly enriched uranium is worth $11,000 to $17,000 when downblended. The 2300 tons of highly enriched uranium in nuclear warheads are worth $26-$39 billion.

2300 tons of highly enriched uranium would supply about seven years total demand for the world's reactors.

5. The impact of reprocessing on disposition

Plutonium can be chemically dissolved and removed from spent reactor fuel.

Reprocessing was a response to rising uranium prices and a shortage of uranium enrichment capacity in the 1970's.

The cost of chemical reprocessing was staggering - the total cost for reprocessing the spent fuel destined for the first US repository would be $126 billion - and by the 1980's the economic environment for reprocessing had changed

The THORP reprocessing plant should yield a rate of return an investment of only about 1.9%. Actual performance of THORP is even below this expectation.

Recent studies show that a closed fuel cycle is less economical than an open fuel cycle with direct disposal. One study showed a 25% advantage for the open cycle.

Reprocessing prices range from $750 to $2000/kg. When reprocessing costs are attributed to the cost of mixed oxide fuel, it is about six times as expensive as low enriched uranium fuel.

Government ownership of reprocessing and transmuting facilities is the only realistic alternative because of the lower capital costs.

To be competitive, the cost of pyroprocessing must be about $315/kg or less, a reduction by a factor of six to seven times of the cost of large scale aqueous reprocessing.

6. Disposition through mixed oxide burning - fabrication and operations issues

Mixed oxide fabrication plants have a capacity of about 88 tons/year. Construction of additional capacity of 260 tons/year is under way in France, Japan and Germany.

The amount of plutonium that can be used in light water reactors is limited to about one third of the core.

Costs for mixed oxide fuel are five to six times those for low enriched uranium reactor fuel.

No commercial reactor operator will accept mixed oxide fuel without either subsidies to compensate for the additional costs or increased charges to the consumers of power.

Government ownership of mixed oxide fuel fabrication facilities saves approximately $600 million in capital costs.

The subsidies required by existing reactor mixed oxide users - after full reimbursement for fuel displacement credits - would be from $1.92 billion to $3.11 billion with government financing and from $2.09 billion to $3.71 billion with private financing.

The subsidy provided to a mixed oxide user must cover not only the difference between the price of mixed oxide fuel and the price of the most economically competitive energy alternative.

7. Burning plutonium in reactors - implications for disposition

If reactor plutonium is stored for long periods it progressively turns into Americium 241 which must be removed before mixed oxide fuel can be fabricated.

Weapon-grade plutonium is contaminated with gallium which must be removed before it can be used in mixed oxide fuel.

The net burn up plutonium for fast breeder reactors is only about 38%.

Plutonium cannot be recycled more than once or twice because plutonium isotope quality decays in the presence of the radiation.

Proponents of breeder reactors acknowledge they are not economical. Only Japan retains interest in breeders.

An advanced light water reactor costs about $4.3 billion of which $1.6 billion is capital cost, $2.3 billion is fixed operating cost, and $.4 billion is incremental operating cost.

An advanced liquid metal reactor costs about $5.6 billion of which $2.2 billion is capital costs, $2.9 billion is fixed operating costs, and $0.5 billion is required for incremental operating costs.

Three options for using plutonium in reactors and their costs are:

1. Use weapon-grade plutonium in existing fast reactors without reprocessing - cost: $19,500/kg.

2. Use light water reactors with one-third or partial mixed oxide fuel without reprocessing - cost: $8,300/kg with weapon-grade plutonium.
3. Use light water reactors with full mixed oxide fuel loads without reprocessing - cost: $6,000/kg with weapon-grade plutonium.

None of these options has any commercial value.

By comparison, storing plutonium for 20 or more years would cost about $4,100/kg and mixing plutonium with waste and dispose of it as waste would only cost about $1,080/kg in marginal costs over storing the waste alone, or about $5,200/kg.

Storage costs raise the cost of burning plutonium in light water reactors by $4400 to $10900/kg.

A new mixed oxide fabrication facility costs $425 million to $1.3 billion and would take about a decade to complete.

Total disposal costs for geologic disposal of spent fuel would range between $100 and $310 million for 50 metric tons of plutonium without vitrification.

8 . A comparison of proposals for disposition of plutonium

Disposition options that appear to be non-viable for technical or economic reasons.

1. Underground fissioning by nuclear explosion - cost: about $2.2 billion for 20,000 pits. These costs increase when underground pollution of aquifers and other resources are considered.
2   Burning plutonium in light water reactors - cannot be economically justified in a commercial power environment.
3. Burning plutonium in fast breeder reactors - not economically feasible.
4. Burning plutonium in unconventional matrices - not significantly different from burning mixed oxide fuel and not economically feasible.
5. Launching plutonium into the sun - cost: about $29 billion to dispose of 250 tons of weapon grade plutonium.
6. Ocean-based disposition - cost: a few hundred million dollars for emplacement with development and demonstration costing billions of dollars. Costs are increased by the likelihood of pollution and the chance of recovery of plutonium.
7   Mix and melt disposition - cost: similar to the $4 billion cost of pyroprocessing. Requires a giant reprocessing facility that has not yet been developed.
8. Transmutation - cost: uncertain but no less than $50 billion and could easily exceed $100 billion. Not possible before 2015.

Disposition options that appear to be viable on both economic and technical grounds.

1. Direct disposal in surface storage - cost: at plutonium storage costs of $4.35/gram/year, storing 50 metric tons of plutonium for a decade would cost from $1 to $2 billion.

2. Mined geologic disposal - cost: estimated at $130-$400 million for 50 metric tons of plutonium without vitrification. In practice costs have been much higher - $1.6 billion had been spent on the Waste Isolation Pilot Project through April, 1995 and Yucca Mountain costs are estimated at $22 billion.

3. Deep borehole disposal - cost: each hole would cost about $110 million and would hold 50 metric tons of plutonium. Development costs would substantially surpass emplacement costs.

4. Vitrification - cost: approximately $1.8 billion for all US surplus weapons plutonium or $4.9 Billion to $5.8 Billion for all scrap plutonium whose purity ranged from 100% to .001%. The total cost of vitrifying and storing plutonium would be about $500,000 per canister - $300,000 for fabrication and $185,000-$200,000 for storage.

1 Economic issues and the disposition of weapon-grade plutonium and highly enriched uranium

Introduction

On February 23, 1941, an experiment at the University of California at Berkeley produced Plutonium 238 by bombarding uranium with deuterium. By the end of March, Pu-239 had been isolated, and by 1944 the first kilogram quantities of Pu were being produced (Seaborg, 1995). Tom Grumbly, US Assistant Secretary of Energy, notes that while Plutonium is named for Pluto, the Roman god of the underworld, the name also comes from *ploutos*, which means wealth. And Grumbly claims "the scientists working on the Manhattan project appreciated the double meaning of Pu." (1995, p. 18)

Meanwhile, the Y-12 plant at Oak Ridge, Tennessee, had begun production in November, 1943. Its theory and design had also originated at the University of California at Berkeley, and early work at the plant was devoted to isolating and producing the isotope of Uranium-235 needed for manufacturing bombs (Oak Ridge Education Project, 1992, p. 7). This isotope of Uranium, enriched to high levels of purity, became known as highly enriched uranium (HEU).

This book explores the economic and cost issues surrounding a major area of expenditure now facing the US: safely disposing (disposition) of the massive amounts of weapon-grade plutonium and highly enriched uranium (HEU) created during the Cold War. Disposition could potentially be accomplished by means that range from transmuting the surplus nuclear materials into other elements, mixing the materials with spent nuclear fuel, vitrifying the materials by mixing them in molten glass, shooting the materials into the sun, or 'burning' the materials for power generation in nuclear reactors.[1]

Disposition and the allocation of national resources

In December, 1996, the DOE announced it was pursuing a 'dual track' approach to disposition involving both vitrification and using plutonium as reactor fuel ('burning' mixed oxide - MOX - fuel in civilian power reactors) (Lippman, 1996, p. A01). These two methods and many others will be discussed in later chapters, but whatever method or methods are finally selected for disposition, they are sure to be very expensive and politically charged.

The mere fact that a disposition method is not economically feasible does not remove it as a potential solution. For example, over the last three years the uneconomical aspects of burning plutonium in reactors in the form of MOX have been made abundantly clear by a number of studies. However, of all the materials, systems, facilities, and laboratories involved in the design and operation of nuclear weapons during the Cold War, the most readily available assets for reuse are often identified as being the HEU and plutonium from warheads. For example, Glenn Seaborg, the physicist most responsible for developing plutonium, claims that:

> Plutonium is the key to the long-term contribution of nuclear power to meeting the world's growing needs for energy. By allowing us to burn virtually all of the available uranium rather than just 1% as we do at the present, the use of plutonium makes nuclear energy by far the largest energy resource available.... (1995).

In spite of an overwhelming body of evidence that there is currently no way to economically use plutonium as reactor fuel, the need to dispose of plutonium and HEU in a safe and environmentally sound manner remains. This problem is, among other things, a fundamental issue of resource allocation. Under the current federal budgeting philosophy in the United States, programs managed by the Department of Energy (DOE) tend to compete with one another for the total funds assigned to that agency. For example, in the FY 1995 DOE budget a tradeoff was made between increased funding for nuclear weapons and reduced funding for nuclear site cleanup.

Similar tradeoffs are likely to be made when disposition is funded, and disposition is likely to compete directly or indirectly with other alternatives for energy funding. For example, if subsidized by the US government, research into plutonium as reactor fuel or the operations associated with its use are likely to consume funds that might otherwise be available to support competing, sustainable energy alternatives or cleanup activities. Further, using plutonium generally implies reprocessing - a chemical process to separate plutonium from spent nuclear reactor fuel or from other substances.

Both the US and Russia have considerable experience with reprocessing to generate defense nuclear materials. US defense-related reprocessing has created about 100 million gallons of high-level waste that contain 99% of all radioactivity in US nuclear wastes. This waste is stored in 243 underground tanks in Washington, South Carolina, Idaho, and New York. Sixty-seven tanks at Hanford, Washington are known or suspected to have leaked high-level waste into the surrounding soil (Closing the Circle on the Spitting of the Atom, 1996, pp. 30, 31). Russia, an advocate of reprocessing, has approximately 66 million gallons of liquid radioactive waste in storage tanks and another 100 billion gallons of radioactive waste in open

2

ponds and special pools (Kushnikov, 1995, p. 27). Cleaning up wastes of this type creates a further demand on the limited resources available for energy-related expenditures.

In spite of this, quasi-private consortia have put considerable effort into convincing the US government to embark on reprocessing spent nuclear fuel to recover plutonium and uranium, and to support plutonium burning. These efforts have either:

1   Assumed there was an economical way to burn plutonium for power,
2   Proposed the construction and operation of new reactors specifically built to burn plutonium as part of a employment-creation program at old nuclear weapon sites, or
3   Claimed that even if power generation itself was uneconomical, it would still provide a way to dispose of the large stocks of plutonium and thus, was worthy of government support.

At the same time, other technical solutions for the plutonium problem have also been proposed. Many of these are transmutation techniques that will require large amounts of federal research and development money to construct facilities to turn plutonium into shorter-lived elements.[2] Others, such as shooting plutonium into the sun, are equally expensive. The potential costs posed by this difficult set of decisions are the subject of the chapters that follow.

Non-proliferation

If disposition of weapon-usable nuclear materials is likely to be so expensive, why should this country, or any nation, be interested in pursuing it as a national objective? When evaluating any disposition option, one should keep in mind that the major obstacle to building a nuclear bomb is getting the fissile material. When that obstacle is overcome, the rest is fairly simple. Plutonium and highly enriched Uranium-235 (HEU) can both be used to make nuclear weapons. But of the two, only HEU can be diluted with naturally occurring uranium (U-238) to make low-enriched uranium (LEU). LEU is the fuel used by most of the world's reactors and it can neither sustain the fast-neutron chain reaction needed for a nuclear explosion nor can it be easily reconverted to weapon-usable material. Re-enriching LEU is a complex process requiring equipment most proliferators do not possess (Management and Disposition of Excess Weapons Plutonium, 1994, p. 4).

On the other hand, plutonium cannot be diluted with other isotopes of plutonium to make it unusable for weapons. According to the National Academy of Sciences (NAS),

separating plutonium from other elements with which it might be mixed in fresh reactor fuel requires only straightforward chemical processing. Thus, the

3

management of plutonium in any form requires greater security than does the management of LEU (1994, pp. 4, 5).

This is particularly true given the fact that while 17 kg of HEU are required to make a bomb, only about 5 kg of weapon-grade plutonium or 7 kg of reactor grade plutonium will make a nuclear weapon (Chow and Solomon, 1993, p. 13).

The world-wide problem of plutonium accumulation has been complicated by the large amount of plutonium generated by civilian nuclear power operations. The NAS notes that plutonium of virtually any isotopic composition can be used to make nuclear weapons. Although reactor-grade plutonium does present some complications in making a nuclear weapon, the difference in proliferation risk between it and weapon-grade Pu is small (1994, p. 4, 5). The Pu-240 in reactor-grade plutonium generates about 1 million neutrons per second through spontaneous fission - thus making premature chain reactions likely and a 'fizzle' yield more probable. However, on July 23, 1945, Robert Oppenheimer stated that the fizzle yield of the 20,000 ton Nagasaki bomb would still be 1000 tons of TNT - more than enough to destroy a large target (von Hipple, 1995, p. 32).

Reactor-grade plutonium has 19% or more Pu-240 and 55-60% Pu-239. Weapon-grade plutonium has 7% or less of Pu-240 and almost 93% Pu-239 (Makhijani and Makhijani, 1995, p. 10). The use of reactor-grade plutonium in a nuclear weapon was successfully demonstrated at the Nevada Test Site in 1962 (Makhijani and Makhijani, 1995, p. x). Because of its higher proportion of Pu-240, gamma and neutron doses are larger for reactor grade plutonium than they are for weapon-grade material. If the Nagasaki bomb had been made from reactor grade plutonium, the dose 1 meter from the bare pit would have been about one rem/hr. This is large - radiation doses for nuclear workers are limited to 5 rem/year - but a short-term lethal dose of radiation is at least a few hundred rem. Therefore, committed terrorists could easily handle reactor-grade material (von Hipple, 1995, p. 32).

Surplus weapon-grade plutonium exists in a great variety of forms that include fabricated metallic weapons components, metal and oxide, various scraps and residues. Many scraps and residues may require treatment to assure long-term chemical stability (Kass and Erickson, 1995, p. 11). Those residues exist in both liquid and solid form, and generally contain less than 28% plutonium. 'Stabilization' prepares this plutonium scrap and residues for storage until final disposition can take place (Grumbly, 1995, p. 17, 19). Of the roughly 50 metric tons (MT) of plutonium declared surplus, roughly one-third requires stabilization to address near-term health and safety concerns (Rudy, 1995, p. 22).

Over the next 10 years about 200 tons of Pu and 1,000 tons of HEU will be recovered from dismantled weapons. By comparison, the current plans for civilian worldwide nuclear development call for the separation of more weapon-usable plutonium from spent fuel by the year 2003 than will be generated from dismantled nuclear weapons. Through the year 2003, about 330 tons of reactor-grade plutonium will be separated from spent fuel. Since only 7 kg of reactor-grade plutonium are required to make a primitive bomb, by 2003 there will be enough

4

surplus plutonium from weapons to make 40,000 primitive bombs and enough reactor-grade plutonium to make 47,000 bombs. After 2003, the amount of surplus weapon-grade plutonium will stay about the same, but by 2010 there will be enough reactor-grade plutonium to make about 71,000 primitive bombs (Chow and Solomon, 1993, pp. xi, xiv, xv). Burning plutonium for power legitimizes both reprocessing of spent fuel and the possession of plutonium, and it could vastly complicate these proliferation issues.

Officials of the International Atomic Energy Agency (IAEA) readily concede that reprocessing technology is creating more nuclear material than they can reliably track. In fact, Japan's reprocessing project, still in its infancy, takes up roughly 20% of the agency's inspection resources (Wald and Gordon, 1994). And modern bomb-making technology has made the IAEA's controls and inspections even more critical. Building a nuclear bomb now takes much less plutonium than was originally required. According to the Natural Resources Defense Council, the threshold level of plutonium from which a bomb can be constructed should be lowered from 8 kg (17.6 LB) to 1 kg (2.2 lb.) (Building A-bombs, 1994)

The goals of disposition

Because concerns about proliferation drive the search for an acceptable method of disposition, the primary threats to any disposition program are assumed to be:

1   Theft of plutonium or HEU by an individual or sub-national group.
2   Diversion of plutonium or HEU to illicit uses by a member of the host nation's own nuclear organization infrastructure, in violation of the international regime and before final disposition has taken place.
3   Retrieval of nuclear materials previously disposed of for the purpose of constructing nuclear weapons.
4   Conversion of nuclear materials after disposition to put those materials back into weapons usable form (Duggan, et al., 1995, p. 415).

To combat these threats, the NAS listed the following goals for the long-term disposition of plutonium:

1   Minimize the time during which the plutonium is stored in forms readily usable for nuclear weapons.
2   Preserve material safeguards and security during the disposition process, maintaining the same high standards of security and accounting applied to stored nuclear weapons.
3   Result in a form from which the plutonium would be as difficult to recover for weapons use as the larger and growing quantity of plutonium in commercial spent fuel. This is known as the 'spent fuel standard.'

4   Meet high standards of protection for public and worker health and for the environment (1994, p. 2).

The NAS felt that the two most promising alternatives for achieving these aims were:

1   Fabrication and use as fuel, without reprocessing, in existing or modified nuclear reactors.
2   Vitrification in combination with high-level radioactive waste.

A third option, burial of the excess plutonium in deep boreholes is also being investigated (1994, p. 2).

The Department of Energy (DOE) has also stated goals for a long-term disposition program. These goals, while similar to those of the National Academy of Sciences, differ in several important respects:

1   Resistance to theft or diversion by unauthorized parties.
2   Resistance to retrieval, extraction and reuse by the host nation.
3   Technical viability.
4   Environmental, safety and health compliance.
5   Cost effectiveness.
6   Timeliness.
7   Fostering progress and cooperation with Russia and other countries.
8   Public and institutional acceptance (Rudy, 1995, p. 23).

Based on these criteria, the DOE developed the following core set of technology options for surplus plutonium disposition:

1   Immobilization options in which plutonium is emplaced in glass, ceramic or glass-bonded zeolite waste forms. Immobilization includes a range of technologies like vitrification, that can meet the spent fuel standard for disposition of surplus plutonium (Grumbly, 1995, p. 17).
2   Various reactor options in which surplus plutonium is fabricated into mixed-oxide fuel (MOX) and used in domestic or Canadian nuclear reactors to generate power.
3   Deep geologic disposal options in which plutonium in an appropriate form would be emplaced in a deep borehole (roughly 2-4 km deep) and sealed for isolation from the accessible environment (Rudy, 1995, p. 23).

The end objective of any disposition program is to meet certain key security objectives. In the view of the DOE, the key concerns are to:

1   Minimize the risk that weapons or fissile materials could be obtained by unauthorized parties.

2   Minimize the risk that weapons or fissile material could be reintroduced into the arsenals from which they came.

3   Strengthen national and international arms control mechanisms (Management and Disposition of Excess Weapons Plutonium, 1994, p. 3).

Whatever option is chosen for disposition, a great deal of further study must be done before any option for disposition is finally selected and, as in all other aspects of this problem, even the selection process will be expensive. The NAS claims that the total expenditures for a study of disposition options similar to one they suggest would be several billion dollars spread over one or more decades (Management and Disposition of Excess Weapons Plutonium, 1994, p. 3). If one assumes a $3 billion budget spread over one decade, the disposition selection process could cost the US $300 million per year.

The negative externalities of wastes generated by disposition

In the past, most cost comparisons between different forms of disposition have neglected to add the costs associated with the wastes generated during the processing of nuclear materials for disposition, the assumption being that these costs are approximately identical no matter what kind of disposition option is undertaken. There is no evidence this assumption is correct, and the costs of the waste generated by each disposition option must be carefully compared - particularly when reprocessing is considered.

In any process where material is put in a reactor, whether for power generation or simply to dispose of material, the volume of material remains constant throughout the time it is in the reactor. The composition of the material changes, and radioactivity may become more intense due to the shorter half-lives of elements that may be created or changed. However, under any circumstances volume will only change minutely and the volume of new fuel essentially determines the volume of spent fuel that is generated.

A commercial light water reactor (LWR) that produces one gigawatt (GW) of energy in a year also accumulates up to 800-1000 kg of highly radioactive fission products and approximately 250 kg of plutonium (Kushnikov, 1995, p, 25). All commercial LWRs generate about 2,000 metric tons of spent fuel each year. DOE estimates that the inventory of spent fuel will be about 61,000 tons by 2010, when a waste repository is scheduled to open. DOE further estimates that when the current generation of US nuclear power reactors has been retired and/or replaced by 2030, these reactors will have generated 90,000 metric tons of spent fuel. The repository authorized by the US Congress has a statutory capacity limit of 70,000 metric tons (Developing Technology to Reduce Radioactive Waste May Take Decades and Be Costly, 1993, pp. 11, 26).

Whatever wastes are created during disposition will be added to the large volume of nuclear waste already located at various sites around the US. A June, 1996

7

estimate by the DOE of the amount of waste that would be generated in different cleanup scenarios at these sites was 1 million to 100 million cubic meters (Closing the Circle on the Spitting of the Atom, 1996, p. 87). If reprocessing is used in the disposition of surplus nuclear materials, and particularly if aqueous (liquid-based) reprocessing is involved, the amounts of radioactive waste are likely to increase significantly.

The time frame for the disposition decision

Disposition can be divided into three distinct, but overlapping, phases: dismantling the weapons and warheads, intermediate disposition of nuclear materials, and long-term disposition of all surplus weapon-usable fissile materials (Gray, et al., 1995, p. 58). At the present time, both the US and Russia are involved in dismantlement activities, and some intermediate disposition has taken place at Pantex in the US and will soon occur at the MAYAK facility in Russia. However, the intermediate and long-term disposition stages will each take substantial amounts of planning, time, and money to successfully complete. Further, storage sites for spent nuclear fuel will not be ready until 2010 at the earliest (Chow and Solomon, 1993, pp. xiv, xv) and disposition technologies like vitrification (mixing nuclear materials with molten glass) could not be implemented by the DOE for fifteen or more years after the order to begin the process was given (Carter, 1994, p. 43). Even burning plutonium in power reactors would require storage for up to 10 years due to reprocessing problems, the need to build suitable reactors, and other factors (Chow and Solomon, 1993, pp. xiv, xv).

The National Academy of Sciences has stated that:

> Since it is crucial that at least one of [their] options [MOX or vitrification] succeed, since time is of the essence, and since the costs of pursuing both in parallel are modest in relation to the security stakes, we recommend that project-oriented activities be initiated on both options, in parallel, at once. (1994, p. 417)

Anyone familiar with the history of duplicative research efforts for either weapons or space projects (for example, the concurrent development of the old Vanguard and Redstone projects) has good reason to be skeptical about the effect of such projects on the costs of disposition. But however the costs of disposition are calculated, it is the costs of waste disposal and the cost of security that are likely to have a significant influence on the final costs of any disposition program. In a 1993 article on disposition, von Hipple et. al. noted that:

> Although disposal of plutonium with radioactive waste would forgo the electricity it could generate, this loss is insignificant in the larger context. At present uranium and plutonium prices, plutonium will not be an economic

fuel for at least several decades. In addition, one or two hundred tons of the metal could power the world's current nuclear capacity for only a fraction of a year. The security threat posed by this material should therefore take precedence (1993, p. 48).

In addition, the reality of this difficult choice is that there are so many known and potential delays already involved in any disposition program that there is sufficient time to make a safe, intelligent and economical choice - and there is no reason to rush to a decision without proper consideration of the costs of all options.

Notes

1   Burning is the techno-slang word for using Pu or HEU in nuclear reactors by down-blending (essentially, diluting) HEU to reactor-strength uranium or mixing Pu with uranium to form a mixed oxide fuel (MOX) that can be burned in light water reactors (LWRs).
2   Elements with half-lives of 50 to 100 years instead of the 24,000 years possessed by plutonium.

References

'Building A-bombs Requires Less Material Than Had Been Believed, Experts Say' (1994), *New York Times News Service*, August 21.
Carter, Luther J. (1994), 'Let's Use It', *The Bulletin of the Atomic Scientists*, Vol. 50, No. 3, p. 43.
Chow, Brian G. and Solomon, Kenneth A. (1993), *Limiting the Spread of Weapon-Usable Fissile Materials*, National Defense Research Institute, RAND, Santa Monica, CA, p. 13.
Closing the Circle on the Spitting of the Atom (1996), US Department of Energy, Office of Environmental Management, pp. 30, 31.
Developing Technology to Reduce Radioactive Waste May Take Decades and Be Costly (1993), GAO/RCED-94-16, United States General Accounting Office, Washington, DC, pp. 11, 26.
Duggan, R.A., Jaeger, C.D., Moore, L.R., Tolk, K.M. (1995), 'Non-Proliferation, Safeguards, and Security for the Fissile Materials Disposition Program Immobilization Alternatives', *Final Proceedings: US Department of Energy Plutonium Stabilization and Immobilization Workshop*, p. 415.
Gray, L., Kan, T., Shaw, H., Armantrout, G. (1995), 'Immobilization Needs and Technology Programs', *Final Proceedings: US Department of Energy Plutonium Stabilization and Immobilization Workshop*, p. 58.

Grumbly, Thomas, P. (1995), 'Plutonium Stabilization and Immobilization Workshop Objectives', *Final Proceedings: US Department of Energy Plutonium Stabilization and Immobilization Workshop*, p. 18.

Kass, Jeffrey N. and Erickson, Randy (1995), 'Workshop Perspectives', *Final Proceedings: US Department of Energy Plutonium Stabilization and Immobilization Workshop*, p. 11.

Kushnikov, Viktor (1995), 'Radioactive Waste Management and Plutonium Recovery Within the Context of the Development of Nuclear Energy in Russia', *Final Proceedings: US Department of Energy Plutonium Stabilization and Immobilization Workshop*, p. 27.

Lippman, Thomas W. (1996), 'US To Burn, Bury Toxic Plutonium From Weapons', *Washington Post*, December 9, p. A01.

Makhijani, Arjun and Makhijani, Annie (1995), *Fissile Materials In A Glass, Darkly*, IEER Press, Takoma Park, Maryland, p. 10.

Management and Disposition of Excess Weapons Plutonium (1994), Committee on International Security and Arms Control, National Academy of Sciences, National Academy Press, Washington, DC, p. 4.

Rudy, Greg (1995), 'Overview of Surplus Weapons Plutonium Disposition', *Final Proceedings: US Department of Energy Plutonium Stabilization and Immobilization Workshop*, p. 22.

Seaborg, Glenn T. (1995), 'Preface', *Protection and Management of Plutonium*, American Nuclear Society Special Report.

Oak Ridge Education Project (1992), *A Citizen's Guide to Oak Ridge*, A Project of the Foundation for Global Sustainability, p. 7.

von Hippel, F., Miller, M., Feiveson, H., Diakov, A., Berkhout, F. (1993), 'Eliminating Nuclear Warheads', *Scientific American*, August, p. 48.

von Hipple, Frank (1995), 'Fissile Material Security In The Post-Cold-War World', *Physics Today*, June, p. 32.

Wald, Matthew L. and Gordon, Michael R. (1994), 'Russia And US Have Different Ideas About Dealing With Surplus Plutonium', *NY Times News Service*, August 19.

2 The economic value of plutonium and highly enriched uranium

Economic value

Value is normally established through a market mechanism in which a buyer and seller negotiate a price viewed as fair by each. Ideally, the negotiated price of the good should reflect all the costs incurred in its production. But if the market mechanism is influenced by outside, non-market forces (such as government intervention or subsidies) the price of the good is altered - and often it is artificially lowered.

After the Second World War, when private nuclear reactors began to produce plutonium in the process of generating energy, the federal government paid nine dollars a gram for plutonium produced in civilian reactors. This price was not set by a market mechanism and it remained at nine dollars a gram until the program was discontinued in 1970 (McPhee, 1973). At the present time there is no legal market for weapon-grade plutonium, due partially to a lack of demand, partially to plutonium's role as a heavily controlled substance, and partially to adverse public reaction over its shipment and use. Japan has purchased non-weapon-grade plutonium from France for future use in its breeder reactor program, but Japan's unique lack of alternative energy sources make its determination of the value of plutonium inapplicable to most other countries. It is probable there is another, illicit market for plutonium, but prices in this market are surely much higher than the real energy value of plutonium. Hence, neither previous experience nor the illicit market provide useful guidance about the actual value of plutonium.

Viewed from an economic perspective, the value of plutonium or HEU is simply the amount of economic return the use of either material would generate in an open market. According to current plans, HEU from the former Soviet Union is to be de-enriched by US Enrichment Corporation (USEC) at its plants in Paducah, Kentucky and Portsmouth, Ohio. USEC is supposed to be a for-profit company, and during these operations a market price for HEU will be established. Whether this price will incorporate the actual costs incurred in downblending HEU is another matter. However, even assuming that the price of HEU is based on its potential energy

11

value absent any government subsidy, there is no similar market mechanism for determining the market value of weapon-grade plutonium.

Plutonium as an asset or liability

With no market to determine prices, one is forced to resort to a second-level analysis in the case of plutonium; namely, whether plutonium is an asset or a liability. Here the distinction is simple. Something is an asset if more economic benefits accrue from its use than costs, and it is a liability if the reverse is true. In this context, the issue is whether using plutonium to generate electrical power would result in a cost savings over the use of more conventional uranium-based fuels [meaning it is an asset] or whether using plutonium in this manner would cost more than conventional nuclear power generation [meaning it is a liability].

One rough way to determine the asset/liability question is to ask whether those who would use the material in a competitive environment - for example, commercial power producers - would accept plutonium if it were free. Richard Garwin evaluated the economics of power generation with both mixed oxide fuel (MOX) and uranium and found that a fuel fabricator will not accept free weapon-grade plutonium for MOX, but will, instead, pay for low enriched uranium (LEU) (1993). This implies that plutonium has a negative value that exceeds the cost of LEU.

This theory was tested in December, 1995, when DOE asked the nation's electric utilities to indicate whether they had an interest in offering one or more reactors for MOX burning. The DOE received positive responses from sixteen power supply systems. These responses were based on the prospect of both free MOX fuel and an additional subsidy from the US government. Given these economic benefits, there may have been more positive responses had the DOE not tied this request to a related option to produce tritium for weapons (Numark, 1996, p. 6). However, even though the free MOX fuel reduced the negative value of plutonium by shifting the costs of MOX fabrication to the US government, a government subsidy was obviously still required to fully offset the remaining negative value of the material and to make it slightly less expensive than LEU.

Both Garwin's calculations and the DOE offer suggest that plutonium is a liability and has a negative value. However, DOE has often cited a positive valuation of weapon-grade plutonium based on the costs required to manufacture the material. DOE's pricing theory appears to be that if something costs a great deal to produce, it must be worth a great deal of money. The fallacy in such an argument is clear, but this has been DOE's principle method of value calculation. Further, DOE's assignment of costs has always been somewhat selective. An actual accounting of the true costs of generating plutonium and HEU through dismantlement of nuclear weapons would include:

1   Research costs accumulated in developing the materials.

2   Initial costs to extract uranium, to purify the materials and to make plutonium in reactors or to generate HEU through gaseous diffusion.
3   The cost to fabricate the materials into weapons.
4   The cost to maintain the materials in weapons.
5   The cost to dismantle the weapons and free the materials for other uses.
6   Costs of waste handling and disposition at each of the above steps.
7   Future costs of permanent disposal and storage.

Garwin has noted that the world can't recover these expenditures and, thus, there will be no offset to what society must pay to cover the environmental damages and cleanup costs of military nuclear programs. Nor is the stock of HEU and plutonium either different from fuels available for power reactors or uniquely valuable (1992, pp. 17-20). Further, the use of past costs to assign value is not a market-based approach, nor are these costs necessarily rational given the manner in which DOE operations are conducted. As a result of all these points, the past costs of generating plutonium are usually counted as sunk costs of the Cold War. Disposition alternatives are evaluated under the assumption that all past costs are sunk and future decisions should be based only on future costs.

In summary, there is no open, operating market in weapon-grade plutonium. Further, there is also a very real threat that introducing the available amounts of HEU into the low enriched uranium market could potentially destroy the uranium industry in the US. As a result, prices for plutonium and the amounts of HEU allowed on the market at any given time have been set by national governments for political, not economic purposes (Bund, 1996). This type of government intervention creates non-economic pricing factors that cascade through all commercial operations using plutonium or HEU. When materials whose supply is artificially constrained or whose market value is not freely determined are introduced into a commercial power-generating regime, careful market analysis and cost control no longer govern which power sources are exploited.

Disposition and economic value

When alternatives are evaluated for disposition, certain physical rules apply:

1   Reactors using any acceptable material - low enriched uranium from virgin ore, plutonium-based MOX, or low-enriched uranium from down-blended HEU - will generate the same energy value (Silvestri, 1994).
2   The total quantity of material put into a reactor will become the total quantity of spent fuel generated by that reactor.

As a result of these rules, only two cost comparisons are necessary to show whether plutonium or HEU are assets or liabilities; i.e., whether they can be used for commercial power generation with an economic benefit:

1   The cost of processing and fabricating reactor fuel - and whether this cost would be higher or lower when plutonium or HEU is used.
2   Whether the cost of disposing of plutonium and highly enriched uranium through other means (vitrification, blending with waste, etc.) and storage of the waste might be lowered by burning them in a reactor, or whether the overall costs of disposition can be reduced by simply disposing of either material without first submitting it to a reactor. Here, there must be counted among the costs those of possible reuse in weapons if the materials are disposed of improperly and the health costs associated with potential worker and public exposure during handling of the materials.

This second issue incorporates proliferation and health risks in addition to economics. US policy clearly mandates that in order of importance, proliferation and health risks come first and economics comes second. On this issue, the National Academy of Science found that:

> Exploiting the energy value of plutonium should not be a central criterion for decision making, both because the cost of fabricating and safeguarding plutonium fuels makes them currently not competitive with cheap and widely available low-enriched uranium fuels, and because whatever economic value this plutonium might represent now or in the future is small by comparison to the security stakes (1994, pp. 3, 4).

Proliferation and health risks are likely to be reduced if disposition consists of as few steps as possible because the risks of proliferation and contamination are directly related to increased handling. For example:

1   Reprocessing, MOX fabrication, and transmutation all require increased handling of plutonium-laden materials.
2   Attempts to reach a spent fuel standard by burning materials in a reactor could be accomplished more simply and with less handling by mixing the plutonium-laden material directly with spent fuel.

Cost comparisons for various disposition options are governed by the following factors:

1   Increased storage of plutonium increases costs.
2   Increased amounts of waste increase disposition costs.
3   Waste storage costs are increased by burning because volume of spent fuel increases.
4   When reprocessing is involved, aqueous reprocessing causes both storage costs and waste volume to increase.

5   Transmutation may lower storage costs by lowering storage time. However, materials that undergo transmutation still require so many years of storage that discount rates make cost comparisons based on storage time irrelevant.

6   Transmutation costs are increased by the need to handle hotter materials.

7   Transmutation costs are increased by the need to reprocess materials that will be submitted for transmutation.

8   Costs are increased by research and development (R&D). Disposition methodologies such as transmutation that still have to undergo the costs of research and development are likely to be more expensive than other options that will not incur R&D costs.

Specific cost considerations for reprocessing during disposition

In the past, it was assumed that if nuclear power from fission is to be employed over the long run, there would be a point at which one would run out of economical natural uranium and shift to artificially-bred isotopes. These isotopes would be recovered by reprocessing the spent fuel of commercial power reactors and separating out plutonium so it could be incorporated in MOX fuel and used again in power reactors. In the commercial market, the shift to MOX would obviously occur at the point at which an economic advantage would accrue from reprocessing spent fuel and burning plutonium (von Hipple, 1995, pp. 2-3).

As the following sections will show, reprocessing only becomes desirable if there is a shortage of natural uranium available at reasonable prices. However, such a shortage is unlikely to occur for a long time - if at all. If 8.4 kg of natural uranium are required to produce 1 kg of LEU, and if recycling uranium and plutonium through reprocessing accounted for 25% of all commercial power needs, an increase in the price of natural uranium to $130/kg from a reference price of $50/kg would increase the relative cost of the LEU used in reactors by only $170/kg (von Hipple, 1995, pp. 2-3). This increase in the price of uranium would make its discovery and mining more profitable, increasing the amount of known uranium deposits and making profitable the exploitation of deposits that had been marginal at lower prices.

The world supply of uranium recoverable at $130/kg has yet to be estimated, but it is believed to exceed 20 million tons. Beyond this, there are huge deposits in Central Asia and Australia that could be exploited economically for $50/kg. Even if one assumes the 'high growth, high nuclear' scenario advanced by the International Panel on Climate Change in 1992, the world's nuclear capacity would only increase to 1250 (gigawatts-electric) GWe in 2050 and 2700 GWe in the year 2100. At these power levels, a once-through light water reactor (LWR) fuel cycle for all the world's reactors would use only 5 million tons of uranium by 2050 and 17 million tons by 2100. 10% of the world's estimated low-cost uranium resources would amount to about 2.1 million tons - an amount that would last until the year 2030. Thus, at the current spot price of about $16 ton, there is clearly no incentive to increase

15

known reserves and, because of its cost, there is also no incentive to engage in reprocessing to recover plutonium from spent fuel (von Hipple, 1995, pp. 2-3).

If there is no economic incentive, why are some countries currently reprocessing spent fuel? There are a number of non-economic reasons for such a decision:

1   Several countries like France and Belgium started reprocessing before it was clear it was not economical.
2   Reprocessing earns hard foreign exchange currency.
3   A country with limited natural resources, such as Japan, may view reprocessing as important to national survival because it reduces dependence on other countries (Albright and Feiveson, 1988, p. 254).
4   Reprocessing may involve subsidies from the government that hide the actual costs of fuel production.
5   Dealing with waste and safeguarding the plutonium, the high cost parts of reprocessing, may not be necessary for the reprocessor if the customer is another country that takes back both the plutonium and the vitrified waste.
6   In Russia, France and the UK, reprocessing maintains large work forces in the same manner as defense production (von Hipple, 1995, pp. 2-3).

Even if plutonium is, by definition, a waste material (or liability), there is one additional reason that has been advanced for bearing the expense of plutonium separation. Plutonium is generally safer to store in a pure form than when mixed with other materials. Plutonium weapon components are essentially pure, and they have been stored safely for decades. A great deal is known about how to prepare and package this kind of plutonium metal for long-term, stabile storage. However, plutonium residues are mixed with many other substances and there is little long-term storage experience with these materials. There has been a general feeling among some in DOE that returning plutonium residues to a pure form would ease storage problems (Hurt, 1995, pp. 36-37).

There are two obvious weaknesses in this reasoning. First, this rationale applies only if one is willing to absorb the increased costs of reprocessing, security and monitoring necessary to store purified plutonium residues for an extended time. A decision to do this would be rational only if the plutonium had some economic value and if the costs of reprocessing, storage and security did not exceed this value. Plutonium's status as an economic liability means that a rational economic decision-maker will not be willing to pay any costs for reprocessing or storage. And second, reprocessing plutonium residues adds to the quantity of purified, weapon-grade plutonium at same time that everyone is trying to reduce the surpluses of plutonium accumulating from warhead dismantlement.

The value of recycled nuclear reactor fuel without reprocessing costs

If the costs of reprocessing are disregarded or not borne by the potential user of plutonium, then some would claim that plutonium is a free good, the use of which might generate sufficient returns to even allow the utility to recover some of the sunk costs it incurred in reprocessing. However, plutonium for which the costs of reprocessing have been written off will only be an economic benefit if the costs of fabricating MOX fuel from the 'free' plutonium are less than those for making LEU.

When calculating the cost of MOX fabrication, the costs of storing the plutonium over the period between acquisition and use become a significant issue. The cost of LEU may be compared with that of comparable MOX fuel in the following manner:

1   At $120/kg for uranium - the mid-range price estimated for European nuclear reactors by the Organization for Economic Cooperation and Development (OECD), based on 2%/yr price escalation - one metric ton of 3.1% LEU would cost about $1.94 million in 1996 dollars. [In 1996, the spot price of uranium was actually $16.30/kg, about $104 less than the assumed OECD price.] Of the $1.94 million price, $1 million is charged for uranium extraction, $690,000 for separative work, and $250,000 for fuel fabrication.

2   One metric ton of comparable MOX would cost about $1 million. Since one metric ton of MOX contains 27 kg of fissile plutonium, the value of the plutonium becomes the amount saved on fuel prices. MOX fuel is $940,000/MT cheaper, so the savings would be about $35/gram of plutonium. These savings are realized if the plutonium is recycled into MOX fuel immediately after its separation from spent fuel. However, based on the European experience, the OECD study assumed an average storage time of five years. At the end of any substantial period of storage, some of the plutonium will have decayed to Americium 241, and this must be removed before MOX can be fabricated. Reasonable storage costs for plutonium for five years, plus the fabricator's charge to remove Americium 241 total about $20/gram. Thus, these charges reduce the value of plutonium to about $15/gram in 1996 dollars (Albright and Feiveson, 1988, p. 253).

The value of recycled nuclear reactor fuel with reprocessing costs

Reprocessing costs alone make the fuel cycle costs for plutonium about 10% higher than those for LEU used on a once through cycle - based on an average uranium price of $50 per kg. The price of uranium would have to reach $330/kg in 1996 dollars before the use of reprocessed plutonium offered any savings (Albright and Feiveson, 1988, p. 253). Further, as the price of uranium drops, the losses incurred when plutonium is used obviously rise.

When both reprocessing and MOX fabrication costs are considered the comparison becomes even worse. At prices approximating 1996 levels of $16/kg for natural uranium and $80/separative work unit (SWU) for separative work, LEU will cost about $790/kg. This is about half the estimated cost of $1450-$1800/kg for new MOX fuel. At these prices, burning plutonium in reactors would add an additional $1.7 billion to the costs to dispose of 200 tons of plutonium, and the total costs of disposition could be $2-$10 billion using a reasonable range of future LEU and MOX prices (Fetter, 1992, pp. 144-148).

The magnitude of the losses likely to be incurred with plutonium use are demonstrated in the OECD study where reprocessing costs are $1000/per kilogram of heavy metal (kgHM) in 1996 dollars. Based on the concentration of fissile plutonium in spent fuel (about .66% by weight), this cost corresponds to a cost to produce separated plutonium of about $150/g of fissile plutonium. This is ten times the $15 value calculated in (b) above for plutonium without considering reprocessing. Further, the uranium recovered in reprocessing increases the value of recycled products so much that at a cost of $120/kg of uranium, about 2/3 of the value of 'recycling plutonium' actually comes from uranium and only 1/3 comes from plutonium (Albright and Feiveson, 1988, pp. 253-254).

For all these reasons, the question of whether plutonium should be treated as an asset or as a liability is neither subjective, nor is it difficult to answer with a fair degree of precision. Virtually all independent studies show that plutonium burning will not be economically feasible for at least the next 50 years because expenses associated with plutonium use are increased due to:

1   Criticality issues that necessitate handling plutonium in small (non-critical) amounts.
2   Increased costs to separate plutonium through reprocessing of spent fuel.
3   Increased costs to store, safeguard, and handle plutonium because of its potential weapon applications and because of criticality issues.
4   Increased costs of waste handling and cleanup created by the additional handling and processing needed to make MOX fuel.
5   High costs for new plants to make and burn MOX fuel, if these plants are built. These costs must be amortized over the life of the plant and passed on to electric consumers.

Since surplus plutonium has no legitimate use other than reactor fuel, it is clear why even the French have decided to give plutonium a zero value in their accounts and Germany and Britain have already zeroed out any economic value for plutonium (New Scientist, 1995). Unfortunately, these accounting actions still do not accurately reflect plutonium's status as a liability - plutonium's value is not zero as the French, German, and British actions would imply. As John Gibbons, the White House Science advisor has stated, "plutonium has essentially a negative economic value" (Wald and Gordon, 1994).

Russian perceptions of the value of plutonium

Russia's approach to valuing plutonium appears, like the DOE's, to be at least partially based on the costs expended to create it. Viktor Mikhailov, Russian Minister of Atomic Energy, has said that plutonium cost the Former Soviet Union six times as much to make as HEU so it is unacceptable to destroy it (Garwin, 1992, pp. 17-20). He has also stated that "we have spent too much money making this material to just mix it with radioactive wastes and bury it." (Wald and Gordon, 1994)

However, many Russians recognize that if plutonium has any real economic value, it is only as a future energy source. The Director of the Obninsk Institute has acknowledged that plutonium does not have any economic value in the near term, but he noted that doesn't mean that the economy won't eventually change to favor the use of "products for future technology, which we cannot use today." Russian engineers are planning to use existing supplies of reactor-grade plutonium long before such use becomes economically viable - possibly to avoid the handling difficulties encountered as components of reactor-grade plutonium break down and become more radioactive over the next few years. This will allow Russia to keep its weapon-grade plutonium in storage for the next few decades. (Wald and Gordon, 1994)

It is probable that the near-term attempts by Russia to use plutonium as reactor fuel are partially based on two perceptions - both of which affect assessments of value:

1   Such use allows more oil and gas to be exported. One ton of weapon-grade plutonium, when burnt in an open fuel cycle thermal reactor, is equal to 2.5 billion cubic meters of natural gas. The annual effect of substitution of gas by nuclear power in 1996 dollars would save 25 billion cubic meters or, at export prices, $2.1 billion. For the entire period, the savings would be 440 billion cubic meters of gas or $37 billion in export prices (Ryabev et al., 1996, pp. 2, 5).

2   Shipping reactor fuel is easier than shipping coal, oil, or gas in a country as large as Russia. Viktor Mikhailov has proposed using part of the $10 billion in hard currency Russia will generate by selling 500 metric tons of highly enriched uranium to the United States to help build new breeder reactors to produce more plutonium. Completing the first two reactors and a MOX factory to turn the plutonium into reactor fuel would cost $2.5 billion, according to experts at the Institute of Physics & Power Engineering (Wald and Gordon, 1994).

Assuming these potential uses of plutonium actually represent an assessment of its value instead of bureaucratic inertia or other rationale, these assessments are not universally accepted in Russia. Aleksei Yablokov, an adviser to President Boris Yeltsin and a former environment minister, claimed it is not clear that existing nuclear reactors, let alone new ones, make economic sense when Russia could replace all its reactors with natural gas and coal-fired power plants for an estimated $6 billion to $7 billion. In contrast, he claimed the International Atomic Energy Agency (IAEA) has stated that upgrading existing Russian reactors to Western safety standards would take between $26 billion and $120 billion (Wald and Gordon, 1994). Estimates of the investment required by the Russian nuclear power industry based on the most optimistic scenario is $8.83 billion. One half of this is for safety enhancements (Ryabev et al., 1996, p. 7).

Additional costs associated with plutonium use and storage

Many of the cost figures this chapter presents for MOX fabrication and plutonium separation through reprocessing assume that the facilities to use fabricate MOX, reprocess spent fuel, and use plutonium-based fuels already exist. Building a system to separate and burn plutonium is neither a trivial nor an inexpensive task. In the Office of Technology Assessment (OTA) of the US Congress, estimated the following resource requirements for a minimum, intermediate and maximum program to use plutonium:

1   A minimum acquisition program. This would be based on a gas-graphite production reactor rated at 30 MW-thermal (MWt) and capable of producing around 8 kg of plutonium per year. This program is estimated to require a capital cost of $130-$325 million in 1996 dollars, of which $38-$110 million is the construction cost of the reactor (Technologies Underlying Weapons of Mass Destruction, 1993, p. 156). The average cost of the reprocessing component is $16-$44 million, or about 12% of the total. The time for construction of this project is estimated to be 3-4 years, with a crew of 100 (Lyman, 1995, p. 448).

2   An intermediate acquisition program. This program is capable of producing around 100 kg of plutonium per year. A 400 MWt reactor was estimated to require a capital investment in 1996 dollars in the range of $435 million to $1.09 billion (for the reactor alone), with a cost overrun of up to 100% possible due to delays. The construction time for this reactor was estimated to be 5-7 years, requiring a staff of 200-300 (Technologies Underlying Weapons of Mass Destruction, 1993, p. 158). The cost of the reprocessing plant was not given; however, scaling up from the previous example at 40% of reactor cost yields a value of $170-$440 million (Lyman, 1995, p. 448).

In times of national emergency, actual construction times could be considerably shorter than these estimates. The 250 MWt B-Reactor at Hanford, the first plutonium production reactor of the Manhattan project, was completed in about a year; 15 months later, two other reactors were in operation. Together, the three reactors were producing plutonium at a rate of about 100 kg/year within two years after the start of construction (Cochran et al., 1987, pp. 59-64). Thousands of workers were employed in these projects.

3   A maximum acquisition program. In this program the production rate is limited only by the resources available. For example, during the Cold War, multiple 2150 MWt reactors were constructed at the Savannah River Site, each capable of producing about 600 kg of plutonium per year. The cost of each reactor of this size was estimated between $1.8 and $4.4 billion in 1996 dollars. The average time between start of construction and startup was under three years (Cochran et al., 1987, p. 61). Again, thousands of workers were required.

In addition to the costs related to separation, MOX fabrication and reactor construction for plutonium burning, there is an additional cost associated with disposing of the waste products generated in this process. If an underground repository is planned, large mining operations typically require capital investments on the order of a few hundred million to well over one billion dollars, based on the isolation of the site and the difficulty of the material which is to be mined (Peters, 1987, p. 262). Development times of 2-5 years before production can begin is normal (Hartman, 1987). The excavation of spent fuel storage sites would require a similar or larger effort and investment. For example, the development costs proposed for the Yucca Mountain repository included $265 million for site preparation, $425 million for construction of shafts and ramps, initial excavations, and underground service systems, and $675 million for construction of surface facilities, for a total of $1.36 billion in 1996 dollars (The Cost of High-Level Waste Disposal: Analysis of Factors Affecting Cost Estimates, 1993, p. 136). These costs have risen as delays have occurred in every aspect of the project and Yucca Mountain is now estimated to cost a total of $22 billion.

The energy value environment

All the costs covered in this chapter occur in an environment where their true relevance can only be determined by comparing them with the costs of other, competing sources of energy. It is at this point that one encounters the 'energy value trap.'

The 'energy value trap' works in the following manner: plutonium will become an economically viable source of energy only when a shortage or some other cause dramatically increases the price of uranium. However, when such a price rise occurs, the ability of all nuclear power plants to compete with other sources of energy is diminished and nuclear power itself becomes less economically feasible.

Energy pricing does not occur in a vacuum - it accounts for all competing means of energy production. Hydro power from dams is already cheaper than nuclear power in the Northwest US. Coal fired plants are still being built based on economic analysis of their returns. Meanwhile, no new nuclear plant has been started in the

21

US since 1978. This means that the very forces that might make plutonium competitive with uranium might also make all nuclear power noncompetitive with other sources of energy. Scenarios that forecast increasing dependence on nuclear power, depend on the depletion of other resources and the assumption that solar power and other renewable sources will not be employed. This is increasingly unlikely based on the experience of the last twenty years.

References

Albright, David, and Feiveson, Harold A. (1988), 'Plutonium Recycling and the Problem of Nuclear Proliferation', *Annual Review of Energy*, Vol. 13, p. 254.

Bund, Matthew (1996), Personal Communication to William Weida, March 17.

Cochran, T., Arkin, W., Norris, R., Hoenig, M. (1987), *Nuclear Weapons Databook, Volume II: US Nuclear Warhead Production*, Ballinger, Cambridge, MA, pp. 59-64.

Fetter, Steve (1992), 'Control and Disposition of Nuclear Weapons Materials', *Working Papers of the International Symposium on Conversion of Nuclear Warheads for Peaceful Purposes*, Rome, Italy, pp. 144-148.

Garwin, Richard L. (1992), 'Steps Toward the Elimination of Almost All Nuclear Warheads', *Working Papers of the International Symposium on Conversion of Nuclear Warheads for Peaceful Purposes*, Rome, Italy, pp. 17-20.

Garwin, Richard L. (1993), *Critical Question: The Value of Plutonium*, Presentation at Cornell University.

Hartman, H. (1987), *Introductory Mining Engineering*, Wiley, New York, NY.

Hurt, David (1995), 'Progress on Plutonium Stabilization', *Final Proceedings: US Department of Energy Plutonium Stabilization and Immobilization Workshop*, pp. 36-37.

Lyman, Edwin S. (1995), 'A Perspective on the Proliferation Risks of Plutonium Mines', *Final Proceedings: US Department of Energy Plutonium Stabilization and Immobilization Workshop*, p. 448.

Management and Disposition of Excess Weapons Plutonium (1994), Committee on International Security and Arms Control, National Academy of Sciences, National Academy Press, Washington, DC, pp. 3, 4.

McPhee, John (1973), *The Curve Of Binding Energy*, The Noonday Press, New York, p. 37.

New Scientist (1995).

Numark, Neil J. (1996), *Get SMART: The Case for a Strategic Materials Reduction Treaty, and Its Implications*, Paper for the International Conference on Military Conversion and Science: Utilization/Disposal of the Excess Weapon Plutonium: Scientific, Technological and Socio-Economic Aspects, Como, Italy, p. 6.

Peters, W. (1987), *Exploration and Mining Geology*, p. 262.

Ryabev, L.D., Favorsky, O., Subbotin, V., Kagramanian, V., Oussanov, V. (1996), *Nuclear Power Strategy of Russia*, Paper for the International Conference on Military Conversion and Science: Utilization/Disposal of the Excess Weapon Plutonium: Scientific, Technological and Socio-Economic Aspects, Como, Italy, pp. 2, 5, 7.

Silvestri, Mario (1994), 'Remarks', *International Congress on Conversion of Nuclear Weapons and underdevelopment: Effective Projects from Italy*, Rome.

Technologies Underlying Weapons of Mass Destruction (1993), Office of Technology Assessment, US Congress, (OTA),OTA-BP-ISC-115, US Government Printing Office, Washington, DC, p. 156.

The Cost of High-Level Waste Disposal: Analysis of Factors Affecting Cost Estimates (1993), Nuclear Energy Agency/Organization for Economic Cooperation and Development (NEA/OECD), OECD, Paris, p. 136.

von Hipple, Frank (1995), *Reprocessing of Spent Power-Reactor Fuel: Why We Can Wait, Why We Should Wait*, International Conference on Evaluation of Emerging Nuclear Fuel Cycle Systems, Versailles, France, pp. 2-3.

Wald, Matthew L. and Gordon, Michael R. (1994), 'Russia And US Have Different Ideas About Dealing With Surplus Plutonium', *NY Times News Service*, August 19.

3 Disposition quantities of surplus materials

Introduction

The surplus of weapon-usable plutonium and HEU comes from two sources:

1. Dismantlement of nuclear weapons by the US and the former Soviet Union.
2. Reprocessed spent fuel from nuclear reactors.

The 'pit' or explosive center of a modern nuclear warhead is composed of about 3-4 kg of plutonium. Most warheads also have a uranium 'secondary' composed of about 15 kg of HEU with over 90% U-235 (von Hippel et al., 1993, p. 46). These materials make up the military surplus weapon-grade plutonium and HEU.

Table 3.1
World-wide totals of separated plutonium

Type	1970	1980	1990	1995	2010
Military	130	210	250	250	250-265
Commercial	5	40	120	141	370-546

Sources: 1970, 1980, 1990, figures: Sachs, Noah (1996), *Risky Relapse into Reprocessing*, Institute for Energy and Environmental Research. 1995 and 2010: figures based on Albright, et. al. (1993), *World Inventory of Plutonium and Highly Enriched Uranium, 1992*, Oxford University Press, pp. 203-206 and Berkhout, Franz (1996), *Briefing to Reprocessing Workshop*, Washington, DC, October 4.

Table 3.1 shows the estimated metric tonnage of the global accumulation of separated plutonium. Commercial nuclear reactors are a growing part of the disposition problem. By the year 2000, the total amount of separated plutonium in the civilian sector is expected to surpass the total amount in weapons arsenals (Sachs, 1996, p. 20). Currently, separation of civilian plutonium in Europe is proceeding at the rate of about 20 tons per year (Berkhout, 1996).

Since only a fraction of the plutonium in spent fuel has been separated, the figures in Table 3.1 significantly understate the total amount of plutonium when non-separated quantities are added. For example, at the end of 1990 there were 532 tons of plutonium in spent fuel world-wide (Albright et al., 1993, p. 199). By 2010, US civilian nuclear power reactors will have produced about 70,000 tons of spent fuel (Nuclear Wastes: Technologies for Separations and Transmutation, 1996, p. ix). If this spent fuel contains an average of about 4% plutonium, there will be 2,800 tons of plutonium from spent fuel alone.

Total quantities of plutonium

In 1991, the US had about 19,000 nuclear warheads and the Former Soviet Union (FSU) had about 35,000 nuclear warheads scattered across the four nuclear republics that came out of the old Soviet Union. Under START I and START II, signed in July, 1991 and January, 1993, the US and the FSU agreed to reduce to 3,500 and 3,000 strategic warheads respectively by 2003. Numbers of remaining tactical warheads may vary, but a good estimate would be about 1,500 US and 2,000 FSU tactical warheads. Thus, each side is scheduled to have about 5000 nuclear warheads in 2003 (von Hippel et al., 1993, pp. 44, 46). About 2,500 warheads could be dismantled each year in the US, but only about 1,170 will be dismantled if parity is maintained with the FSU's rate of about 2,250 per year (Chow and Solomon, 1993, pp. 9, 10).

US supplies of plutonium

The supply and locations of plutonium available for reuse in the United States as of September, 1994, are shown in Table 3.2. Aside from these inventories, the total amount of plutonium available for use in non-weapon applications is directly dependent on dismantlement of old warheads. Since the pit of a nuclear weapon is composed of 3-4 kilograms of plutonium and about 15 kg of HEU, surplus US warheads contain about 50 tons of plutonium and up to 400 tons of HEU (von Hippel et al., 1993, pp. 46-47).

Table 3.2
US plutonium inventories and locations

Site	Plutonium Inventory (in Metric Tons)	Plutonium in Wastes (in Kilograms)
DOD & Pantex	66.1	N/A
Rocky Flats	12.7	47
Hanford	11.0	1,522
Argonne Lab-West	4.0	2
Los Alamos	2.7	610
Savannah River	2.0	575
INEL	.5	1,106
Lawrence Livermore	.3	N/A
Others	.2	N/A
Oak Ridge	N/A	41
Nevada Test Site	N/A	16
TOTAL	99.5	3,919

Source: O'Leary, Hazel (1996), *Plutonium: The First 50 Years - United States Plutonium Production, Acquisition, and Utilization from 1944 to 1994*, Department of Energy, Washington, DC, February 6.

US warheads are dismantled at the Pantex Plant in Amarillo, Texas. An explosive layer around the pit is first removed in an explosion-proof cell so the pit itself is revealed. Tritium, a gas used to boost the explosive output of the warhead, is then removed and returned to the Savannah River Site in South Carolina for recycling. The pits are stored in W.W.II era concrete bunkers at the Pantex Plant where they are carefully spaced to avoid criticality problems (Makhijani et al., 1996, pp. 199-201). Over 18,100 warheads were completely dismantled at Pantex between 1980 and April, 1994 (June 27 Openness Press Conference Fact Sheets, 1994, pp. 172-173).[1]

Once the HEU, plutonium and tritium are removed, the shell of the bomb must also be disassembled. This can be done by dipping the shell in liquid nitrogen and dropping a heavy weight on it to shatter the sensitive components. Gold, copper, platinum and silver are recovered from these pieces - and some components are worth as much as $11,000 per ton. These separated components, when sold, may return as much as $1 million to the DOE budget (Holden, 1993, p. 1673 and DOE Realizes Post-Cold War Dividend While Meeting 'START' Targets, 1996).

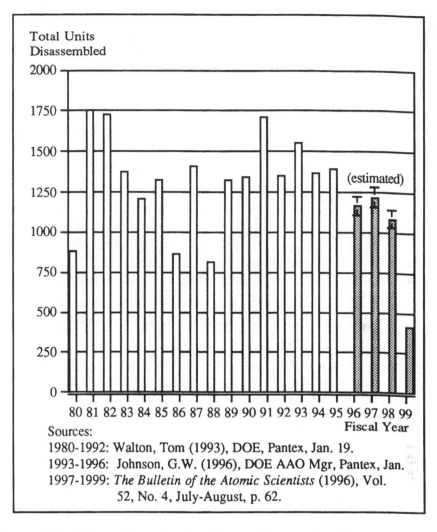

Figure 3.1 Dismantlement rates for nuclear warheads
at the Pantex plant, Amarillo, Texas

Beginning in October, 1992 the DOE set a dismantlement goal of 2000 warheads per year for the US. As Figure 3.1 shows, this goal has never been met and, at the current pace, dismantlement work at Pantex is likely to continue for three or four more years. On January 19, 1994 the DOE announced a decision to increase pit storage at Pantex from 6,000 to 12,000 pits. As of May, 1995 there were 7239 pits stored at Pantex and there were about 8500 pits stored there in May, 1996. Given the remaining storage space - 3500 - it is unlikely there is sufficient space for the remaining warheads scheduled for dismantlement. This will increase pressures to find some other means of disposing of these materials (Norris and Arkin, 1995, pp. 78-79).

Site-specific costs of dismantlement activities in the US are not yet public, but these activities comprised the following parts of the Pantex budget over three years of increasing dismantlement activity:

Table 3.3
Pantex dismantlement as a percent of total workload

Year	Percent of Workload
1988	34
1989	42
1990	62.5

Source: *Letter to Steven Schwartz from Gloria E. Inlow* (1990), Deputy Director, Office of Intergovernmental and External Affairs, US Department Of Energy, Albuquerque Operations Office, November 21.

Since FY1991, DOE has spent a total of about $771 million a year for dismantlement at Pantex and at the Y-12 plant at Oak Ridge. These costs are likely to remain constant through 2003. According to current estimates, storage of nuclear weapon components will continue for two decades or more. The security this storage requires will cost from several billion dollars to more than $10 billion, with costs rising if the storage period is extended (Makhijani et al., 1996, pp. 199-201).

Russian supplies of plutonium

100 or more metric tons of excess weapon grade plutonium are now available in Russia (Kushnikov, 1995, p. 25). Soviet warheads also contained more than 500 tons of HEU (von Hippel et al., 1993, pp. 46-47). According to Anatoli Diakov, the FSU and later, Russia, had produced 126.2 MT of weapon-grade plutonium by 1996. In addition, 72 MT of plutonium was produced by civilian reactors. At present, about 30 tons of plutonium dioxide is stored at Chelyabinsk-65. In addition, the US is helping Russia build another storage facility for dismantled plutonium at MAYAK. Completion of this facility is scheduled for 1997 at a cost estimated at $150 million, of which the US has contributed $75 million in financing (Diakov, 1996). As of 1996, the US had provided a total of about $1.3 billion to Russia to help dismantle nuclear weapons and safeguard nuclear materials (Makhijani et al., 1996, p. 205). Russia also has spent fuel containing approximately 25 tons of plutonium from the VVER 100 and the RBMK reactors. An additional 30 tons of fuel-grade plutonium exist in the form of purified oxide (Kushnikov, 1995, p. 25).

Status of materials in other countries

Most research assumes the supply of surplus fissile materials for which disposition is necessary is confined to Russia and the US. However, France, Belgium and Great Britain, and, to a lesser extent, Germany, have been actively reprocessing spent fuel and generating plutonium for years. These activities will be covered in the reprocessing and MOX sections of this book, but these countries are usually assumed to only produce plutonium that will be converted into MOX and then burned in commercial reactors. Given the nuclear weapons possessed by Great Britain and France, this assumption is clearly erroneous, but their supplies of plutonium have been regarded as technically not surplus and hence, not candidates for disposition (aside from burning in commercial reactors). Another troubling situation - the case of Japan - deserves closer inspection. The Japanese have recently accumulated what appears to be a significant surplus of plutonium through purchases from the French, they have not signed the nuclear non-proliferation treaty, and they are not considering disposition options other than reactor burning. Japanese inventory figures are presented in the following section.

Japanese supply of surplus plutonium

The outlook for Japanese plutonium inventories has changed considerably with the recent accident at Monju. Table 3.4 shows the Japanese plutonium inventory at the end of 1994. According to the long run supply and demand balance first presented by the Japanese Atomic Energy Commission in June, 1994 and revised in August, 1995 the annual supply of plutonium from the Tokai reprocessing plant is about 0.4 ton. This was short of the expected consumption of 0.6 ton by Monju, Fugen and Joyo, and this shortage was expected to be offset by plutonium imports from France. The Japanese government now says that plutonium stored for use in Monju will be used instead in Joyo, but the expected annual demand by Joyo and Fugen together is less than 0.2 tons. Thus, the total Japanese plutonium surplus in Japan and Europe could amount to around 25 tons by the turn of the century (Takagi, 1996).

Table 3.4
Japan's separated plutonium inventory - end of 1994

Facility	Kilograms Of Plutonium	Stockpiled(s) Or In Use /Ready For Use(u)*
Reprocessing plant	836	
As nitrate	710	s
Stored as oxide	126	s
Fuel Fabrication plant	3,018	
Stored as oxide	2,032	s
Under test or processing	948	u
Completed fuel	38	u
Reactor sites	498	
Joyo	6	u
Monju	15	u
Fugen	53	u
Critical assemblies	425	u
Overseas reprocessors	8,720	
UK(BNFL)	1,412	s
France(COGEMA)	7,308	s
Total	13,072	11.588(s)+1.484(u)

*Attribution to u and s by Dr. Takagi.
Source: Takagi, Jinzaburo (1996), Citizens' Nuclear Information Center, 1-59-14-302, Higashi-nakano, Nakano-ku, Tokyo 164, Japan, January 10.

Total quantities of HEU

As opposed to plutonium, HEU is neither used in nor made in reactors. There are about 2300 metric tons of HEU worldwide, almost all of it in the former Soviet Union and the US (Makhijani and Makhijani, 1995, pp. 16-17). Total US HEU production from 1945 to 1992 was 994 metric tons. Of this, 483 metric tons were made at the K-25 facility at the Oak Ridge Site between 1945 and 1964, and 511 metric tons were made at the Portsmouth, Ohio plant between 1956 and 1992 (O'Leary, 1994).

The amount of HEU consumed by the US since 1945 is estimated to be about 105 metric tons including uranium burned in reactors for plutonium production at SRS (about 42 metric tons), uranium burned by the Navy (about 12 metric tons), uranium consumed in research (about 25 metric tons), uranium exported to France and UK (abut 6 metric tons), and uranium consumed in weapons tests (about 20 metric tons). This leaves 630 metric tons [994 - (105 + 259)] unaccounted for in the revealed inventories. This 'missing' amount is probably split between the Pantex stockpile and the remaining nuclear arsenal (Gray, 1994).

Table 3.5
The US inventory of HEU - locations and amounts

Metric Tons	Location	Metric Tons	Location
0.6	Hanford, WA	26.2	INEL, ID
0.2	LLNL, CA	6.7	Rocky Flats, CO
3.2	LANL, NM	0.9	SNL, NM
Classified	Pantex, TX	1.6	Knolls, NY
0.2	Brookhaven, IL	23.0	Portsmouth, OH
168.9	Y-12, SRS, SC	1.5	K-25, ORNL, TN
1.4	ORNL, TN	24.4	SRS, SC

TOTAL = 258.8 metric tons (not including Pantex)

Source: O'Leary, Hazel (1994), Remarks Concerning a DOE fact sheet on HEU, DOE, Washington, DC, June 27.

When the number of nuclear weapons peaked at 32,500, independent experts estimated there were 500-550 metric tons of HEU in warheads, implying about 16 kg per weapon. The amount of HEU per weapon is thought to have declined slightly since then due to greater use of plutonium. However, new estimates suggest that about 50% more HEU was devoted to weapons than previously believed. Thus, either more was used in each bomb than had been estimated - which suggests that about 10 tons more would also have been consumed in tests - or there was considerable overproduction and stockpiling for an arsenal buildup that never occurred (Gray, 1994).

Note

1   Additional weapons were disassembled and then reassembled or modified.

References

Albright, D., Berkhout, F.,Walker, W. (1993), *World Inventory of Plutonium and Highly Enriched Uranium, 1992*, Oxford University Press, p. 199.

31

Berkhout, Franz (1996), *Briefing to Reprocessing Workshop*, Washington, DC, October 4.

Chow, Brian G. and Solomon, Kenneth A. (1993), *Limiting the Spread of Weapon-Usable Fissile Materials*, National Defense Research Institute, RAND, Santa Monica, CA, pp. 9, 10.

Diakov, Anatoli S. (1996), *Utilization of Already Separated Plutonium in Russia: Consideration of Short- and Long-Term Options*, Paper Presented at the International Conference on Military Conversion and Science: Utilization/Disposal of the Excess Weapon Plutonium: Scientific, Technological and Socio-Economic Aspects, Como, Italy.

DOE Realizes Post-Cold War Dividend While Meeting 'START' Targets (1996), US Department of Energy Press Release, April 18.

Gray, Peter (1994), Personal Communication to William Weida, June 30.

Holden, Constance (1993), 'Breaking Up (a Bomb) Is Hard To Do', *Science*, September 24, p. 1673.

June 27 Openness Press Conference Fact Sheets (1994), US Department of Energy, pp. 172-173.

Kushnikov, Viktor (1995), 'Radioactive Waste Management and Plutonium Recovery Within the Context of the Development of Nuclear Energy in Russia', *Final Proceedings: US Department of Energy Plutonium Stabilization and Immobilization Workshop*, p,. 25.

Makhijani, Arjun and Makhijani, Annie (1995), *Fissile Materials In A Glass, Darkly*, IEER Press, Takoma Park, Maryland, pp. 16-17.

Makhijani, A., Schwartz, S.I., Norris, R.S. (1997), 'Retirement and Dismantlement of Nuclear Weapons, and Storage and Disposition of Retired Nuclear Weapons and Surplus Nuclear Weapons Materials', in Schwartz, Steven I., Ed., *Atomic Audit*, Brookings, Washington, DC, pending publication, pp. 199-201.

Norris, Robert S. and Arkin, William M. (1995), 'US Nuclear Weapons Stockpile, July 1995', *The Bulletin of the Atomic Scientists*, Vol. 51, No. 4, pp. 78-79.

Nuclear Wastes: Technologies for Separations and Transmutation (1996), Committee on Separations Technology and Transmutation Systems, Board on Radioactive Waste Management, Commission on Geosciences, Environment, and Resources, National Research Council, National Academy Press, Washington, DC, p. ix.

O'Leary, Hazel (1994), *Remarks Concerning a DOE Fact Sheet on HEU*, DOE, Washington, DC, June 27.

Sachs, Noah (1996), *Risky Relapse into Reprocessing*, Institute for Energy and Environmental Research, p. 20.

Takagi, Jinzaburo (1996), Citizens' Nuclear Information Center, 1-59-14-302 Higashi-nakano, Nakano-ku, Tokyo 164, Japan.

von Hippel, F., Miller, M., Feiveson, H., Diakov, A., Berkhout, F. (1993), 'Eliminating Nuclear Warheads', *Scientific American*, August, pp. 44-47.

4 The economic nature of the US commercial nuclear industry and the use of downblended HEU as reactor fuel

Classical economics and the commercial nuclear industry

The ability of the nuclear industry to produce at a competitive price determines the resulting demand for nuclear power. As the previous chapter stated, energy prices are not set in a vacuum and the 'energy value trap' awaits those products whose use depends on increasing the cost of only one of a set of competing energy alternatives. To continue to supply power to their customers, pricing in the commercial nuclear industry must be competitive with the prices of alternative fuels. In turn, the demand for nuclear power creates a demand for uranium that determines not only its present price, but also the amount of exploration for uranium and hence, the reserves of uranium available to be exploited at various prices. Thus, the proven reserves of uranium that can or will be exploited are directly dependent on the price of uranium.

This would all be a simple economic problem were it not for intrusions of the US federal government into the energy markets in the form of both regulations and subsidies. Since its inception, subsidies have been a way of life in the US nuclear power industry. A 1992 report found that over the period 1950 to 1990, 20% or $110 billion of the $595 billion (in 1996 dollars) spent to develop and obtain nuclear power was provided by the US federal government (Komanoff Energy Associates, 1992). According to the DOE, of total subsidies to the energy sector provided by the federal government in 1992, nuclear energy received $899 million of $4.88 billion expended - or about 18%. However, while most other sources of energy (oil, coal, etc.) received either tax subsidies to lower prices or direct subsidies to encourage consumer use - both of which acted to stimulate demand for the product - nuclear energy received almost all of its subsidies ($890 out of $899 million) in Research and Development. In fact nuclear energy received 44% of all US energy R&D subsidies in 1992 (Energy Information Administration, 1992, p. 7).

In 1993, nuclear energy accounted for about 28% of gross electrical energy generated in the United States. In France and Belgium - both major proponents of reprocessing and MOX use - it accounted for 78 and 60%, respectively, of all electricity generated (Energy Statistics Yearbook, 1995). These differences, and the

33

availability of alternative methods of power generation, are largely responsible for the significant differences in attitude toward nuclear power in general and toward reprocessing and MOX fuel use in particular.

In the period before 1978, the heavy emphasis on nuclear research had an unintended consequence that affected the economic viability of the entire US nuclear industry. The emphasis on research resulted in a multitude of new reactor designs, and the use of many different designs typified the construction of US commercial nuclear reactors. The failure to recognize the advantages of a common reactor design (an advantage exploited by the French) complicated regulatory problems and created economic inefficiencies across the commercial nuclear industry.

Over the last twenty years, funding of nuclear energy research has continued, but with little actual implementation of the results of that research. As construction of new reactors in the US ceased - the last US commercial nuclear reactor was started in 1978 - a few large companies were able to stay in the reactor research and development business. However, government support for large reactor research programs replaced the need to answer the demands of the market with viable commercial products. Each new research proposal was further removed from the last reactor private industry and the public were actually willing to fund and build. New reactors will only be built if they can successfully compete with other forms of energy production. But the US has unwittingly created a nuclear industry interested in the development of sources of power, not the economics of producing power.

The move into research by the US's nuclear reactor builders was also aided by the tightened regulatory environment in the US. New federal regulations attempted to deal with dangers posed by nuclear plants after the Three Mile Island accident and other safety incidents raised public awareness of problems in the US nuclear industry. A stricter regulatory environment made the permitting process for new plants so difficult, and extended the construction time so much, that the economic viability of new commercial reactors was adversely affected.

These factors all help explain the nuclear industry's continuing research into the use of plutonium burning reactors in the face of overwhelming evidence that such reactors would not be economically viable. Further, they explain why, as time has passed, the economic viability of standard nuclear reactors has also deteriorated. This is unlikely to improve in the future when plans to generate power from plutonium are proposed to take place. Evaluating the current status of nuclear power in the US, Shearson Lehman reported that: "Evidence suggests the average operating costs of nuclear power plants are now higher than those of conventional plants and other power supply alternatives." (Electric Utilities Commentary, 1992, p. i) And Moody's investor's service has stated that:

> Given increasing competition from other types of generating facilities and renewed efforts via conservation and demand side management programs to reduce the need for new capacity additions, nuclear power's economics must be comparable with alternative fuel sources and energy efficiency and conservation options. In a deregulating environment, the pressure to maintain

competitively low rates will compel utilities to select the most economic option. And given the challenges outlined above, we do not think that nuclear plants are likely to provide such economic benefits (Nuclear Power, 1993, p. 7).

This view of the financial viability of the commercial nuclear industry is also shared by at least one pro-nuclear industry group. A panel of the American Nuclear Society chaired by Dr. Glen Seaborg found that:

In several of the locations where nuclear power previously enjoyed an economic advantage...the construction of new nuclear plants is not currently competitive with electrical energy supply from other sources...as a result, nuclear power expansion has slowed, and we cannot say with certainty that nuclear power will maintain its current share of electrical energy generation in the near term (Protection and Management of Plutonium, 1995, p. 11).

For years, utilities have had a perverse economic incentive to overbuild generating capacity. Companies were allowed a certain percentage of profit on their assets and the more assets they could accumulate, the better their profits. However, as deregulation of the power industry ends in 1997, energy prices are expected to fall by about 20% over the next five years. Utilities with old nuclear plants will have to allocate the costs of these facilities over an ever-tightening market. Nuclear power, with its costly plants and expensive operating procedures is a clear loser in this scenario (Einhorn, 1996).

The price of uranium

The questionable ability of commercial nuclear reactors to compete with other power generators in the US was directly affected by increases in the price of uranium. Many uranium producers ceased operations after the demand for uranium - and hence, the price of uranium - fell in the 1980s due to the cessation of new reactor construction. Uranium prices that had peaked at $40 per pound in 1979 had fallen to $5 per pound by 1991. When demand finally caught up with supply in this now down-sized industry, and when the supply of down-blended Russian uranium available to western markets was reduced to preserve what remained of the uranium industry, uranium prices began to rise. Between January 1, 1996, and August 31, 1996, the spot price of uranium rose 30% to $16.30 a pound, and prices have risen by 75% since the start of 1995. This is likely to further increase the competitiveness problems in the commercial nuclear industry - problems that started in 1991 when uranium was at historically low prices. Cheap natural gas and deregulation of the power industry will make many reactors candidates for retirement over the next few years (Heinz, 1996, Nuclear Fuel, 1996, p. 2).

Burning plutonium in commercial reactors

Chow and Solomon estimate that thermal cycle plutonium use in commercial reactors will not be feasible until the price of uranium-bearing yellowcake reaches $110/LB in 1996 dollars and they estimate this will not occur for 50 years. They further project that fast reactors will not be profitable until yellowcake prices reach $240/LB ($1996) in about 100 years (1993, pp. xvi, xvii). These calculations appear to agree with a 1995 assessment of the American Nuclear Society which claimed that:

> While significant technical problems and high costs have been experienced in the developmental breeders built to date, it has been demonstrated that the technology is viable. Extensive additional development and demonstration will be required, however, before costs that are economically competitive with fossil fuels and current reactor types can be assured (Protection and Management of Plutonium, p. 12).

The costs of burning plutonium are usually compared with the costs of burning highly enriched uranium (HEU) or low enriched uranium (LEU) in reactors. The inherent costs of regulation, operating efficiency, waste disposal, and contamination associated with nuclear operations are approximately the same for both uranium and plutonium operations and, as a result, these costs are seldom discussed. However, these costs - as well as the increasing price of uranium - are all factors when a country, city, or power supplier considers the price of alternative nuclear and non-nuclear power sources. And in this type of comparison, the very factor that makes the cost of using plutonium compare favorably with the cost of using uranium - a sharp increase in the price of uranium - also makes it more likely that neither fuel will be competitive with non-nuclear methods of power generation.

Downblending highly enriched uranium for disposition as reactor fuel

World uranium production has been about 30,000 tons/year for the last decade. Natural uranium contains .711% U-235 (i.e., it is .711 'enriched'). Low enriched uranium (LEU) has enrichment levels of 3-5% U-235 while weapon-grade highly enriched uranium (HEU) typically contains over 90% U-235, so 30,000 tons of natural uranium equate to 230 tons of HEU or to about 6000 tons of LEU (Schulze, 1992, pp. 65-74). Unlike plutonium, HEU is neither used nor made in reactors. Instead, it is enriched through isotope separation and concentration in large centrifuges or by the gaseous diffusion method currently employed at the Paducah, Kentucky and Portsmouth, Ohio plants of the US Enrichment Corporation.

DOE's 1995 *Draft Environmental Impact Statement on the Disposition of Highly Enriched Uranium* defines HEU as any uranium that is enriched above 20% U-235.

This report assumes an average enrichment of 50% U-235. In 1996 there were about 2300 metric tons of HEU, almost all of it in the former Soviet Union and the US (Makhijani and Makhijani, 1995, p. 16-17).

The mechanics of downblending

'Downblending' is a procedure to de-enrich highly enriched uranium to make it useable as reactor fuel. To render HEU useless for weapons, it can be downblended with large quantities of U-238 (99.3% of natural uranium is U-238). Reconstituting HEU requires difficult isotope separation techniques available in only a few countries. Thus, downblended uranium is fairly proliferation-resistant (von Hippel, et al., 1993, pp. 47).

Weapon grade HEU can be diluted from its 93-95% U-235 composition with natural uranium (.7% U-235), slightly enriched uranium (.8 to 2% U-235), or depleted uranium (.2-.3% U-235) to get the 3-4% concentration required by the reactors. For example, if HEU is diluted with enrichment 'tails' of depleted uranium (DU), 21-34 kg of DU would be required for each kg of HEU, depending on the enrichment level of the HEU and on the enrichment level desired in the final product (Garwin, 1992, pp. 17-20). As a fuel fabrication technology, downblending is price-competitive with fuel made from natural uranium and the economics of downblending are much more favorable than those for fabricating plutonium into MOX (Makhijani and Makhijani, 1995, p. 17).

After HEU is downblended, its use in existing reactors poses no technical problems. The amount of blend stock required for final blending down of 500 tons 93.5% HEU are shown in Table 4.1.

Table 4.1
Blending requirements and yields - HEU

Blend Stock	HEU MT	+Blend MT	=4.4% HEU MT
Depleted U(0.2% U-235)	500	+10,600	=11,100
Natural U(.711% U-235)	500	+12,100	=12,600
Slightly Enriched U(1.5% U-235)	500	+15,400	=15,900

Source: Makhijani, Arjun, and Makhijani, Annie (1995), *Fissile Materials In A Glass, Darkly*, IEER Press, Takoma Park, Maryland, p. 76.

The economics of downblending

If a significant amount of HEU was disposed of by down-blending, the resulting amount of nuclear fuel would lower demand for other US-fabricated LEU. Since utilities have long term contracts for yellowcake and enrichment, releasing all the uranium from dismantled nuclear weapons over a period of 10 years could significantly disrupt the LEU market and would probably result in layoffs and facility closings (Chow and Solomon, 1993., pp. 10,11). In fact, it is easy to imagine a scenario where domestic uranium operations were put entirely out of business if large amounts of HEU are down-blended in an economical manner.

To produce one kilogram of LEU by conventional methods requires the purchase of 9.002 kg of natural uranium - 19.8 LB of U_3O_8 at about $16.50 per pound. Since uranium is an internationally-produced commodity, the price for U_3O_8 is highly variable, depending on the supply and demand for reactor fuel. For example, the spot price for yellowcake was $8.00 per pound in 1992, $5.00 in 1994, and $16.50 per pound in 1996 (Wald and Gordon, 1994). Conversion of U_3O_8 to UF_6 costs about $4.50 per kg. Thus, before transportation costs are added, an investment of about $41 in each kg of natural uranium is required before the UF_6 is delivered to the enrichment facility. This equates to about $367 for the 9.002 kg of natural uranium required to make 1 kg of 4% LEU. In 1996, enrichment charges added about $623 for 6.554 separative work units (SWU) - about $95/SWU. Consequently, 4% assay LEU represents an investment of about $990/kg LEU (Campbell, and Snider, 1996, p. 7).

Downblending HEU to make LEU avoids a number of these costs and may have several environmental advantages over normal LEU production - it prevents the accumulation of tailings and it saves energy (Makhijani and Makhijani, 1995, p. 6). As a result, the use of HEU may have lower externalities during the fuel fabrication phase. For example, 200 metric tons (MT) of HEU would produce about 1,200 billion kWh and result in saving 32,000 MT of natural uranium and 19 million SWUs. Russian HEU could generate 4800 billion kWh and replace 150,000 MT of natural uranium and 90 million SWUs. Globally, there is a potential to replace over three years of natural uranium mining and enrichment (Rougeau, 1996).

The cost of downblending and, by implication, the potential economic return from the sale of downblended fuel depends both on the method of downblending selected and on the concentration of U-235 in the final downblended material. For example, if HEU is blended in molten metal to form .9% enriched uranium which would make it essentially unusable for anything, the cost would be about $13,900/kg in 1996 dollars. Each kilogram of HEU must be diluted with 70 kg of depleted uranium (DU) to make .9% enrichment and disposition of 200 tons of HEU would require 14,000 tons of DU. Thus, the costs to downblend 200 tons of HEU to an unusable level of .9% U-235 level would be about $3.4 billion. Comparable costs using a uranyl nitrate methodology would be almost twice as much - about $22,900/kg (Campbell and Snider, 1996).

If, instead, one wishes to downblend to a 4% LEU level using UF_6 blending, the costs of downblending fall to $3,200/kg of HEU. To reach the same 4% concentration using uranyl nitrate would cost about $5, 200/kg HEU. Since 15 kg of LEU reactor fuel result from each kg of HEU, the potential sales revenue in 1996 from downblending based on the price of LEU fabricated from natural uranium would be about $14,850/kg of HEU (15 kg x $990) (Campbell and Snider, 1996).

38

This revenue far exceeds the costs of either method of downblending to a 4% concentration, but it would only apply if the amount of downblended fuel released to the market was small enough not to alter the price of LEU fuel.

When disposition options are considered, the costs of downblending must be compared with the costs of treating HEU as waste and submitting it to direct disposal. Waste disposal costs are estimated to be about $20 per cubic foot. Each HEU waste storage canister could hold 15 kg of HEU and a 55 gallon drum could hold six canisters. Disposition of 200 tons of HEU would therefore generate 16,700 tons of low-level waste which would correspond to 186,000 drums or 1.3 million cubic feet of disposal volume (Campbell and Snider, 1996, p. 5). Based on these calculations, it would cost $26 million to dispose of 200 tons of material packaged in this manner.

Table 4.2
Cost summary of HEU disposition alternatives

Tons of HEU Dispositioned To Waste Disposal	Tons of HEU Dispositioned To Nuclear Fuel	Net Receipts/(Costs) $Million/Disposition
200	0	($3,400)
150	50	($2,230-$2,100)
70	130	($640-$320)
30	170	$340-$770

Source: Campbell, Ronald L., Snider, J. David (1996), *Cost Comparison for Highly Enriched Uranium Disposition Alternatives*, Y/ES-122, Nuclear Materials Disposition Program Office, Y-12 Defense Programs, Lockheed Martin Energy Systems, Inc., April, p. vii.

Thus, depending on market conditions and the likely prices for natural uranium and separative work, one kilogram of HEU will be worth between $11,000 to $17,000 ($1996) for the foreseeable future when downblended to make LEU. This would make the 2300 tons of HEU in the nuclear arsenal worth $26-$39 billion ($1996). If this amount of HEU was released as downblended reactor fuel over a period of 15 years, its sales would comprise about 15% of the total market (Fetter, 1992, pp. 144-148).

The move toward burning downblended HEU has already been slowed to avoid damage to the uranium market (Bund, 1996). This has complicated the deal negotiated by the Bush Administration and signed by the Clinton Administration in 1993, in which Russia agreed to sell a total of 500 tons of bomb-grade uranium from dismantled warheads - 40% of its stockpile - for $11.9 Billion over the next twenty years if certain conditions are met. The US will later resell the material to fulfill demand for nuclear fuel in domestic and world markets (Management and Disposition of Excess Weapons Plutonium, 1994, p. 5).

Diluted HEU is to be transferred to the US over the 20 year period at an annual rate of 10 to 30 MT per year. The first delivery occurred in the spring of 1995, and the 20 year period of delivery was selected to avoid damage to US uranium enrichment industry (Rougeau, 1996). Originally, the Russian HEU was to have been de-enriched by US Enrichment Corporation (USEC) at its plants in Paducah, Kentucky and Portsmouth, Ohio. However, Russia decided to do the downblending before the uranium was transferred to the US. An additional 400 of the 500 tons of US HEU will also become surplus in the near term and this material will probably need to be diluted at the Portsmouth and Paducah facilities (von Hippel, et al., 1993, pp. 47).

The approximate value of 500 tons of HEU from either Russia or the US can be estimated in the manner shown in Table 4.3.

Table 4.3
The value of 500 MT of downblended HEU

500 tons of HEU = 500 X 94%/4% = 11,750 tons of LEU.
To make this amount of LEU from natural uranium would take
  11,750 tons LEU X 7.436 = 87,373 tons of natural uranium.
  Where 7.436 kgs of feed uranium produce 1 kg of 4% LEU.
In 1996, the value of 87,373 tons of natural uranium at $36.00 kg was
  about $3.14 billion.
The value of enriching work for 1 kg of uranium is 6.554 SWU.
  Thus, the enrichment value of 12,000 tons is
  12,000 tons X 1000 kg/t X 6.544 SWU/kg X 95$/SWU = $74.6 B.
  Where $95 was the price of enriching 1 SWU in 1996.
The value of uranium from 500 MT of HEU generated by dismantling
  either US or Russian warheads is approximately:
  $3.14 billion + $74.6 billion = $78 billion.

Source: Uematsu, Kunihiko (1992), *Comments on 'Weapon Grade Uranium and the Nuclear Fuel Market'*, Working Papers of the International Symposium on conversion of Nuclear Warheads for Peaceful Purposes, Rome, Italy, June 15,16,17, pp. 116-118.

US Enrichment Corporation (USEC) is a quasi-private corporation established to purchase the Portsmouth, OH, and Paducah, KY, enrichment plants from the DOE for the purpose of pursuing down-blending as a private commercial venture. DOE has acknowledged that USEC will market this reactor fuel internationally. The US would not control the spent fuel generated by foreign reactors and this spent fuel would be a candidate for reprocessing to extract the plutonium. No protocols forbid reprocessing or require the return to the US of spent fuel generated from this material (Draft Environmental Impact Statement on the Disposition of Highly Enriched Uranium, 1995).

Based on the February, 1993, agreement between the US and Russia, LEU from 6.1 tons of HEU was to be transferred to the US in 1995 and LEU from 12 tons of HEU was to be transferred in 1996 (Diakov, 1996). This should have resulted in the transfer of about 270 tons of LEU over the two year period. However, by October, 1996, the US Enrichment Corporation (USEC) had taken delivery of only 13 tons of reactor fuel derived from Russian HEU and it turned down repeated requests in 1996 to buy additional down-blended material from the Russians. This reluctance to acquire downblended uranium from Russia is probably a result of USEC's inability to compete economically with LEU made from excess HEU. When USEC sells the regular, low enriched reactor fuel it makes from virgin American uranium, it makes considerably more money than it does when sells Russian fuel, and demand for USEC's American uranium may shrink with added Russian supplies. Thus, from a purely commercial perspective, it is to USEC's economic advantage to limit purchases of Russian LEU (Passell, 1996).

Downblending options

Weapon-grade HEU typically contains over 90% U-235 that must be diluted to levels of 3-5% to generate the low enriched uranium used in reactors (Makhijani and Makhijani, 1995, p. 16-17). As of January, 1996, DOE had declared 165 metric tons of US HEU 'surplus' to the stockpile. Any strategy to down-blend HEU and sell it as reactor fuel will require eventual storage of the highly toxic and radioactive spent fuel - which will still contain both plutonium and HEU (Draft Environmental Impact Statement on the Disposition of Highly Enriched Uranium, 1995).

Four down-blending scenarios have been considered by DOE to meet its stated goals of nonproliferation and realizing the "peaceful beneficial use" of HEU in a way that will return money to the US Treasury (Draft Environmental Impact Statement on the Disposition of Highly Enriched Uranium, 1995):

1   Down-blend to less than 1% U-235 and dispose of as low level waste. This would address all proliferation concerns.
2   Limited commercial use - down-blend 35% of HEU into reactor fuel, the rest to less than 1% U-235.
3   Substantial commercial use - down-blend 65% into reactor fuel, the rest to less than 1% U-235.
4   Maximum commercial use - down-blend 85% into reactor fuel, the rest to less than 1% U-235.

DOE's preferred option is to maximize commercial use which, DOE claims, will return the most money to the US Treasury. However, DOE's Draft Environmental Impact Statement on this issue did not present a credible analysis demonstrating a positive economic return. Further, the maximum commercial use option would create more than 5 million pounds of spent nuclear fuel (2,380 metric tons, assuming an assay of 50% enrichment for 170 metric tons of material) that would have to be disposed of at considerable cost. Under its fastest down-blending scenario - downblend to 4% and sell as reactor fuel - DOE's plan would take 10 years to process 200 tons of HEU. During that 10 years, it is likely that more HEU will be declared surplus. DOE argues this will not increase the amount of spent fuel, since reactors will burn something anyway. In addition, DOE claims the use of downblended HEU will reduce environmental impacts since new uranium will not have to be mined for reactor fuel (Draft Environmental Impact Statement on the Disposition of Highly Enriched Uranium, 1995). For this claim to be true, the use of down-blended HEU will have to be complete enough to shut down the US uranium mining industry. If this occurs, it is questionable whether this industry could ever be restarted.

Another option, downblending to 4% for storage until economic and reprocessing concerns are addressed, has been rejected by DOE who claims it provides "no proliferation advantage over down-blending and selling." However, downblending to 4% and storing retains the fuel option while maintaining security of the material in a relatively stable state containing neither plutonium or HEU (Draft Environmental Impact Statement on the Disposition of Highly Enriched Uranium, 1995).

Table 4.4
Energy available from 1000 tons
of HEU used as reactor fuel in 2005

Expected number of operating plants, 1/1/2005:	493	
Corresponding Electrical Power:	Approx. 400 GWe	
Electrical Power From Plants w/Enriched Fuel:	Approx. 374 GWe	
Electrical Power From LWRs:	Approx. 350 GWe	
Fraction of Power From LWRs:	87%	

Typical Core Characteristics of Large LWRs:	PWR 900MWe	BWR 1GWe
Thermal Power, MW	2775	2894
Uranium Mass in Core	72.5	113.8
Power Density, W/CM^3	104.5	52.4
Specific Power, W/GU	38.3	25.4
Avg. Linear Power Density, W/CM	177	188.5
Refueling Strategy	Annual	Annual
Annual Load of Fresh Fuel (Core Fraction)	1/3	1/4
Average Refueling Enrichment, w%	3.25	2.78
Discharge Burnup (GWd/tU)	33	28
Net Plant Efficiency	.33	.339

Amount of Excess HEU from Weapons:	Approx. 1000 tons

Electrical Energy From Warhead Uranium:

Uranium used in blending	Natural
Equivalent number of 1GWE LWR Plants:	270 PWR & 83 BWR
Electrical energy generated by fuel load during its residence in Core (KWh)	6.4×10^9
Mass of weapon uranium in the annual load of fresh fuel (Kg/plant):	665 PWR & 638 BWR
Number of reloads obtainable from warhead uranium blended with natural uranium	1160 PWR & 358 BWR
Feed time of operating reactors using warhead uranium with load factor of 65%/year	6 years
Electrical Energy obtainable from warhead uranium blended with natural uranium	9.7×10^{12} kWh

Source: Adinolfi, Roberto (1992), *The Burning of Uranium Originated by Nuclear Disarmament in Nuclear Power Stations*, Working Papers of the International Symposium on Conversion of Nuclear Warheads for Peaceful Purposes, Rome, Italy, pp. 56-62.

Burning downblended HEU in light water reactors

Because LEU from downblended HEU is essentially identical to that fabricated from natural uranium, downblended HEU can be burned in all existing LWRs. A typical LWR uses about one ton of U-235 to produce 1000 MWe (1 GWe). The world's 400 reactors would thus produce about 300 GWe from 300 tons of U-235. At weapon-grade enrichment levels, 1000 tons of HEU would yield 930 tons of pure U-235, or about three years total demand for the world's reactors (Garwin, 1992, pp. 17-20).

The specific calculations for the energy potential available by using 1000 tons of excess weapon-grade HEU in the world's reactors in the year 2005 are shown in Table 4.4 (Adinolfi, 1992, pp. 56-62). As this table demonstrates, significant potential energy is available from HEU.

Summary - disposition of HEU as reactor fuel

The use of downblended HEU in reactors makes far more economic sense than the use of MOX. In fact, given the current practices of the USEC, it is apparent that downblended HEU reactor fuel purchased from Russia competes well with domestically fabricated LEU. As a result of this situation, several points become clear:

1   There is no economic rationale for using any MOX fuels until all HEU is disposed of.
2   Downblended HEU has the potential to significantly disrupt the LEU fuel market. As a result, the current practice of the US that allows the same company (USEC) to control both LEU and downblended HEU markets is highly questionable and is unlikely to operate to the benefit of consumers of nuclear fuel.
3   The viability of downblended HEU as a commercial product is still directly linked to the viability of the nuclear industry as a whole. This is particularly true since the competitiveness of downblended HEU fuel depends directly on the existence of higher prices for LEU fuel. Again, the factors that make HEU competitive may not make a difference if the industry itself becomes non-competitive with other sources of energy.

Decontamination and decommissioning costs

Unlike other forms of power production, the nuclear power industry accumulates very significant decontamination and decommissioning (D&D) costs at each plant. These costs must be paid when the plant leaves service and they are supposed to be covered by escrow accounts set aside during the operation of the plants. However, the inability of utilities to accurately forecast either the exact time at which a nuclear reactor will go off line or the costs for storage of contaminated materials at that time often result in insufficient D&D funds being set aside.

From 1975 to 1995, the Nuclear Regulatory Commission sponsored studies to estimate the costs of decontamination for boiling water reactors (BWRs) and for pressurized water reactors (PWRs). These costs were estimated at $145 million ($1996) for PWRs and $190 million ($1996) for BWRs. These estimates assume the use of government-paid low-level waste storage and management at Hanford and price regulation. When prices are deregulated, the prices of decontamination could rise by $100 million ($1996). Real-world D&D costs have been higher than those forecast by the NRC. For example, the cost of D&D of the Trojan plant in 1986 was $250 million ($1996) and the estimated D&D cost for the Connecticut Yankee plant announced in 1996 was $400 billion (All Things Considered, 1996). The original construction cost of Trojan was $1.44 billion ($1996) and thus, D&D could cost about 17.5% of the original construction cost (Rothwell, 1996, p. 18-19).

Notes

1   These figures significantly understate the current estimates of the costs to bury nuclear wastes and decommission reactors.

2   Reprocessing plutonium and uranium from spent fuel and using plutonium-bearing mixed-oxide (MOX) fuel in thermal nuclear power plants.

References

Adinolfi, Roberto (1992), 'The Burning of Uranium Originated by Nuclear Disarmament in Nuclear Power Stations', *Working Papers of the International Symposium on conversion of Nuclear Warheads for Peaceful Purposes*, Rome, Italy, June 15,16,17, pp. 56-62.

All Things Considered (1996), National Public Radio, December 5.

Bund, Matthew (1996), Conversation with William Weida at the International Conference on Military Conversion and Science: Utilization/Disposal of the Excess Weapon Plutonium: Scientific, Technological and Socio-Economic Aspects, Como, Italy.

Campbell, Ronald L. and Snider, J. David (1996), *Cost Comparison for Highly Enriched Uranium Disposition Alternatives*, Y/ES-122, Nuclear Materials Disposition Program Office, Y-12 Defense Programs, Lockheed Martin Energy Systems, Inc., p. 7.

Chow, Brian G. and Solomon, Kenneth A. (1993), *Limiting the Spread of Weapon-Usable Fissile Materials*, National Defense Research Institute, RAND, Santa Monica, CA, pp. xvi, xvii.

Diakov, Anatoli S. (1996), *Utilization of Already Separated Plutonium in Russia: Consideration of Short- and Long-Term Options*, Paper Presented at the International Conference on Military Conversion and Science: Utilization/Disposal of the Excess Weapon Plutonium: Scientific, Technological and Socio-Economic Aspects, Como, Italy.

Draft Environmental Impact Statement on the Disposition of Highly Enriched Uranium (1995), US Department of Energy, Office of Fissile Materials Disposition, Washington, DC.

Einhorn, Bruce (1996), 'Electricity, The Power Shift Ahead', *Business Week*, December 2, pp. 78-80.

Electric Utilities Commentary (1992), 'Are Older Nuclear Plants Still Economic?, Insights from a Lehman Brothers Research Conference', vol. 2, no. 21, p. i.

Energy Information Administration, *Federal Energy Subsidies: Direct and Indirect Interventions in Energy Markets* (1992), SR/EMEU/92-02, US Department of Energy, Washington, DC, p. 7.

Energy Statistics Yearbook: 1993 (1995), United Nations, New York, NY.

Fetter, Steve (1992), 'Control and Disposition of Nuclear Weapons Materials', *Working Papers of the International Symposium on conversion of Nuclear Warheads for Peaceful Purposes*, Rome, Italy, pp. 144-148.

Garwin, Richard L. (1992), 'Steps Toward the Elimination of Almost All Nuclear Warheads', *Working Papers of the International Symposium on conversion of Nuclear Warheads for Peaceful Purposes*, Rome, Italy, pp. 17-20.

Heinz, Mark (1996), 'Uranium Prices Rise on Scarcity, Steady Demand', *The Wall Street Journal*, February 26.

Komanoff Energy Associates (1992), *Fiscal Fission: The Economic Failure of Nuclear Power*, 270 Lafayette, Suite 400, New York, NY.

Makhijani, Arjun, and Makhijani, Annie (1995), *Fissile Materials In A Glass, Darkly*, IEER Press, Takoma Park, Maryland, pp. 16-17.

Management and Disposition of Excess Weapons Plutonium (1994), Committee on International Security and Arms Control, National Academy of Sciences, National Academy Press, Washington, DC, p. 5.

Nuclear Fuel (1996), September 9, p. 2.

Nuclear Power (1993), Moody's Special Comment, Moody's Investors Service, New York, NY, April, p. 7.

Passell, Peter (1996), 'Critics say materiel could fall into terrorists', *The New York Times*, August 28.

Protection and Management of Plutonium (1995), American Nuclear Society Special Report, p. 11.

Rothwell, Geoffrey (1996), *Economic Assumptions for Evaluating Reactor-Related Options for Managing Plutonium*, Paper Presented at the International Conference on Military Conversion and Science: Utilization/Disposal of the Excess Weapon Plutonium: Scientific, Technological and Socio-Economic Aspects, Como, Italy, pp. 18-19.

Rougeau, Jean-Pierre (1996), *A Clever Use of Ex-Weapons Material*, Paper Presented at the International Conference on Military Conversion and Science: Utilization/Disposal of the Excess Weapon Plutonium: Scientific, Technological and Socio-Economic Aspects, Como, Italy.

Schulze, Joachim (1992), 'Burning of Plutonium in Light Water Reactors (MOX Fuel Elements) Compared To Other Treatment', *Working Papers of the International Symposium on conversion of Nuclear Warheads for Peaceful Purposes*, Rome, Italy, pp. 65-74.

von Hippel, F., Miller, M., Feiveson, H., Diakov, A., Berkhout, F. (1993), 'Eliminating Nuclear Warheads', *Scientific American*, August, pp. 47.

Wald, Matthew L. (1996), 'Agency To Pursue 2 Plans To Shrink Plutonium Supply', *The New York Times*, December 10.

Wald, Matthew L. and Gordon, Michael R. (1994), 'Russia And US Have Different Ideas About Dealing With Surplus Plutonium', *NY Times News Service*, August 19.

5 The impact of reprocessing on disposition

Introduction

In 1995, a nuclear industry panel chaired by Glenn Seaborg claimed that:

> ...misunderstandings have led some to propose that plutonium should be treated as a dangerous waste and buried, but this suggestion is rarely if ever accompanied by the explanation that burial does not get rid of it and that plutonium can be eliminated, should this ever prove to be desirable, only by burning or transformation to another element through irradiation with neutrons (Seaborg, 1995).

The report went on to note that while large-scale commercial reprocessing is a costly and difficult operation that only a few highly industrialized countries can undertake, small-scale reprocessing is within the reach of a number of moderately industrialized countries, especially if safety standards are relaxed or ignored. Thus, according to this panel, using the 'spent-fuel standard' for storing plutonium does not adequately guard against national proliferation and consideration should be given to implementing a breeder reactor program along with the reprocessing that implies (Protection and Management of Plutonium, 1995, p. 1, 2, 10).[1]

One year prior to the release of this report, the National Academy of Sciences' *Management and Disposition of Excess Weapon's Grade Plutonium* listed only two promising alternatives for plutonium disposition:

1.   Fabricate as fuel - without reprocessing - for use in existing or modified nuclear reactors.
2.   Vitrify in combination with high-level radioactive waste (Management and Disposition of Excess Weapons Plutonium, 1994, p. 2).

Neither of the NAS's alternatives would lead one to reprocess as recommended by the nuclear industry. The reason for the differences between these positions is largely based on the economics of breeder reactors, reprocessing, and MOX fabrication. Reprocessing is explored in this chapter. Breeder reactors and MOX will be discussed in later chapters.

Reprocessing[2]

Reprocessing is the method by which spent nuclear fuel is separated into its various components (plutonium, uranium, and other fission products). This separation occurs in the rough proportion of 1 part plutonium, 95 parts uranium, and 4 parts other fission products (Berkhout, 1993, p. 2). The US's initial efforts to develop reprocessing technologies was a response to the economic conditions of the times. As the nuclear industry entered the 1970's, its rapid growth had contributed to rising uranium prices and a potential shortage of uranium enrichment capacity. The reprocessing costs projected by the nuclear industry were low, and by 1972, enrichment capacity for fuel made from virgin uranium was fully booked. As a result, the Atomic Energy Commission (AEC) stopped taking orders for virgin fuel fabrication and there was a general consensus within the nuclear community that spent fuel from nuclear reactors should be reprocessed by chemically separating the fuel to recover uranium and plutonium for light water reactor (LWR) use (Nuclear Wastes: Technologies for Separations and Transmutation, 1996, p. 413).

Plutonium is much easier to separate than uranium because plutonium is a different isotope and thus, it can be chemically dissolved and removed from spent fuel. The ease of accomplishing this has led to many of the proliferation concerns for plutonium, and the massive amounts of liquid waste generated by chemical reprocessing have created much of the waste problem now present at nuclear sites. A second, experimental technique called pyroprocessing may be able to separate plutonium through an electro-refining process.

Four general techniques are available for uranium isotope separation: gaseous diffusion, centrifuge, calutrons, and laser isotope separation. Gaseous diffusion plants require the largest investment, infrastructure, and power input. Centrifuges require less power per separative work unit (SWU) and can be on a smaller scale. The technology of Laser Isotope Separation has not been implemented. The Office of Technology Assessment has estimated the cost of a 60,000 SWU/yr centrifuge plant to be $300-$600 million in 1996 dollars (Nuclear Proliferation and Safeguards, 1977, pp. 140, 177, 180).

Reprocessing was expected to reduce total fuel cycle costs. In the late 1960's, estimates of reprocessing costs had unit prices below $85/kg of heavy metal ($1996). In fact, initial estimates of the cost premium required for MOX fuel fabrication were only 20% higher than the costs of fabricating virgin fuel. Further, spent fuel would theoretically be reprocessed within six months of discharge, and returned to the reactor with minimum delay - avoiding storage costs and accruing a significant cost savings.

As a result of these favorable cost estimates, three reprocessing plants were constructed in the late 1960's: a commercial reprocessing plant for civilian use operated at West Valley, New York from 1966 to 1972 when it was closed because of technical difficulties and contamination of ground water offsite. A General Electric (GE) plant at Morris, Illinois, was completed in 1968, and an Allied-General Nuclear Services plant at Barnwell, South Carolina was finished in 1974. Neither of these plants ever entered commercial service (Nuclear Wastes: Technologies for Separations and Transmutation, 1996, pp. 413-414).

By the mid-1970's, optimism about nuclear power had begun to wane. In 1977, the Carter Administration canceled breeder commercialization and plutonium recycling, leaving the Barnwell plant without a license. From 1977 to 1979, the US joined other nations in the International Fuel Cycle Evaluation, and it signed the Non Proliferation Treaty (NPT) in 1978.

The US ban on reprocessing domestic fuels was not joined by France, Germany, Belgium and Great Britain, nor was it joined by the Former Soviet Union, all of whom continued to develop their reprocessing and MOX fabrication programs (Nuclear Wastes: Technologies for Separations and Transmutation, 1996, pp. 414-415). And even though it had banned the reprocessing of domestic nuclear fuel, the US itself was accumulating considerable experience reprocessing defense nuclear materials. The US had reprocessed 99.94% of the spent fuel the DOE and its predecessor agencies regarded as requiring reprocessing. This spent fuel contained about 200,000 metric tons of heavy metal. According to DOE calculations in 1996, this left about 150 tons remaining to be reprocessed (Ford, 1996). As of 1996, the vast amounts of liquid chemicals used in US reprocessing have created 99% of all radioactivity in nuclear wastes in the US and about 100 million gallons of high-level liquid waste are stored in underground tanks in Washington, South Carolina, Idaho, and New York (Closing the Circle on the Spitting of the Atom, 1996, pp. 30, 31).

In the 1980's, the US lifted the ban on domestic fuel reprocessing. However, by this time the favorable economic environment for nuclear energy no longer existed, in part because the economics of reprocessing had changed markedly (Nuclear Wastes: Technologies for Separations and Transmutation, 1996, p. 415). The economic responses of the nuclear industry to the unfavorable economic climate for nuclear energy were evidenced in the following ways:

1. Many new plant orders were canceled (the last US commercial reactor project was started in 1978) and the price of uranium concentrate fell from a peak of $44/LB of uranium hexaflouride in 1979 to $17/LB in three years.
2. New discoveries in the 1970's and 1980's significantly increased the world reserves of uranium and lowered the price of uranium.
3. New gas centrifuge technology and the lack of orders for new plants changed enrichment capacity from a shortage to an oversupply.
4. The Nuclear Waste Policy Act of 1982 gave no economic incentive for vitrification and made disposal of spent fuel in a geologic repository the DOE policy. As would be expected, this shifted the emphasis of waste disposal toward burial and away from more technical options.

50

5. The capital cost of breeder reactors relative to LWRs was higher than forecast in the 1970's and the unit cost of both MOX and liquid metal reactor fuel also increased relative to LWR fuel. As a result, deployment of breeder reactors was delayed indefinitely.

6. The estimated cost to construct and operate commercial reprocessing plants rose sharply due to new regulations and industry recognition that this was a high-risk venture - thereby increasing financing costs (Nuclear Wastes: Technologies for Separations and Transmutation, 1996, p. 415).

These changes not only destroyed any economic impetus for reprocessing or for a closed light water reactor fuel cycle in the 1980's, they continue to this day to make reprocessing and MOX fabrication questionable economic activities.

Table 5.1
Euphemisms for reprocessing

DOE Term	Technical Definition
Actinide Recycling	Reprocessing
Aqueous Processing	Reprocessing
Chemical Processing	Reprocessing
Chemical Separation	Reprocessing
Chemical Stabilization	Reprocessing
Electrometallurgical Processing	Reprocessing
Electrorefining	Reprocessing
Fuel Conditioning	Reprocessing
Materials Stabilization	Reprocessing
Partitioning	Reprocessing
Plutonium Purification	Reprocessing
Processing	Reprocessing
Process Q	Reprocessing
Pyrochemical Separation	Reprocessing
Pyroprocessing	Reprocessing
Stabilization	Reprocessing
Treatment	Reprocessing

Source: *Science for Democratic Action* (1996), Institute for Energy and Environmental Research, Winter, 1996, and DOE Briefing by John Ford, Office of Nuclear Materials and Facility Stabilization, DOE, October 4.

In addition to these economic issues, the political and institutional environment for nuclear energy was also altered. Safety issues (including the near meltdown of the Three Mile Island facility in 1979 and the Chernobyl accident in 1986), waste issues associated with the rapidly accumulating amounts of liquid nuclear wastes and spent nuclear fuel, and proliferation issues (including India's nuclear explosion in 1974 using reprocessed weapons materials from an Indian reactor), all changed the climate for reprocessing and for nuclear power.

As a result of all these factors, reprocessing has become such a politically sensitive issue that the DOE has even developed a large list of euphemisms to avoid using the word "reprocessing" in briefings and publications for public consumption. Table 5.1 provides a list of these terms and their technical translation.

There are two current arguments for reprocessing spent fuel that do not consider using the reprocessed fuel in power reactors. Instead, these arguments concentrate on preparing spent fuel for eventual storage. These arguments are:

1. Reprocessing is necessary to get spent fuel ready for geologic disposal: interim storage would allow DOE to gain more information prior to committing to a storage site and form. In this light, DOE is currently investigating:
   a. Chopping up spent fuel and adding depleted uranium (DU).
   b. Melting spent fuel with neutron poisons.
   c. Dissolving spent fuel in acid and adding DU or a neutron poison.
   d. Glass material oxidation and dissolution system.
2. Reprocessing is necessary to remediate short-term safety problems and environmental problems stemming from corroding spent fuel (Sachs, 1996, pp. 25-31).

While the corrosion argument might be justified in the short-run, it could also imply a long term policy of purposely letting spent fuel corrode and then reprocessing it "out of necessity". Further, by October, 1996, the DOE was researching the possibility that corroding spent fuel could be safely stored in dry casks without reprocessing (Ford, 1996). However, decisions in this area are likely to be made without the benefit of good cost information because cost data are highly uncertain and contradictory. The Foreign Research Reactor Environmental Impact Statement (EIS) claims that life-cycle costs for reprocessing and dry storage are about equal if repository emplacement of HEU spent fuel is found to be feasible (Draft Environmental Impact Statement on a Proposed Nuclear Weapons Nonproliferation Policy Concerning Foreign Research Reactor Spent Nuclear Fuel, 1995, pp. 2-16). On the other hand, DOE's own data show that interim dry storage poses fewer safety, environmental and health risks than reprocessing (Sachs, 1996, p. S-5). Hence, interim dry storage is likely to be less expensive than reprocessing if all costs are considered.

As a further consideration, the National Research Council found that:

New waste forms specifically tailored to retain...fission products much longer within the repository offer the potential for improved performance. Benefits could result regardless of transmutation. Candidate waste forms suggested in earlier studies are silver iodide, compounds of separated technetium and separated neptunium, each in thick ferrous containers; and pollucite for cesium (Nuclear Wastes: Technologies for Separations and Transmutation, 1996, p. 335).

Based on these discoveries and on the unfavorable cost performance of present reprocessing operations, in 1996 the National Research Council concluded that:

> The cost of chemical processing solely for waste improvement (i.e., with no recycle of actinides or fission products to reactors) would be staggering. If a US commercial reprocessing plant were to operate with a unit cost of $2000/kgHM (mostly uranium) in the spent fuel, with no return for recycling uranium and plutonium fuel, the total cost for reprocessing the 63,000 Mg of spent fuel destined for the first US repository would be $126 billion (Nuclear Wastes: Technologies for Separations and Transmutation, 1996, p. 335).

European and Japanese experience with reprocessing costs

Reprocessing occupies an unusual position among industrial processes with respect to the accumulated knowledge of its costs - after thirty years of development and operating experience, cost estimates for reprocessing are still highly uncertain and cost estimates still fail to provide any economic rationale for reprocessing activities. This is partially because of government involvement in all large scale reprocessing operations, partially because of secrecy surrounding the separation of material that is potentially weapon-grade, and partially because its uneconomical aspects have led to a significant reluctance on the parts of both the US and the Europeans to reveal exactly how much reprocessing has cost.

The US decision not to participate in commercial reprocessing has meant that all US studies of reprocessing are only estimates based on multiple assumptions and, rarely, some information about European operations. On the other hand, several European plants have accumulated years of reprocessing experience, and the costs of these operations comprise the only reliable estimates of the actual costs of reprocessing and MOX fabrication. Table 5.2 shows the current and anticipated reprocessing capacities of the European plants and of two other plants in Japan and Russia. This remainder of this chapter will present the information presently available on European reprocessing costs. This will be followed by a section on US-generated reprocessing cost estimates - estimates that are regarded as biased and highly unreliable by most researchers not directly connected to the nuclear industry.

The best examples of modern commercial reprocessing are represented by the THORP plant in Great Britain and the UP3 and UP2 plants in France. In addition, Japan is currently constructing the Rokkashomura plant. These four plants are the basis for most reprocessing cost estimates. A general description of these facilities and their costs follows.

Table 5.2
World oxide fuel reprocessing capacity

Plant	Commissioning Date	Capacity (MTHM)
PNC, Tokai-mura, Japan	Operating	90
MAYAK, Chelyabinsk, Russia	Operating	250
UP3, LaHague, France	Operating	800
THORP, Sellafield, GB	1993	700
UP2-800, La Hague, France	1994	800
JNFS, Rokkasho-mura, Japan	2002?	800
PREFRE, Tapur, India	Operating	150
Total Capacity		3590

Sources: Berkhout, F., Diakov, A., Feiveson, H., Hunt, H., Miller, M., and von Hipple, F. (1992), 'Disposition of Separated Plutonium', *Science and Global Security*, 3, pp. 1-52 and Berkhout, Franz (1993), *Fuel Reprocessing At THORP: Profitability and Public Liabilities*, Center for Energy and Environmental Studies, Princeton University, Princeton, NJ, p. 1.

France

Two plants, UP3 (800 MTHM/yr) and UP2 (400 MTHM/yr) have been designed, constructed, and operated. The UP3 plant, the most recent French plant at La Hague, has been in operation since 1990, and COGEMA has reported a capital cost of about $7 billion in 1996 dollars and design and construction requirements of 25 million man-hours of engineering and 56 million man-hours of field construction (Nuclear Wastes: Technologies for Separations and Transmutation, 1996, pp. 418-419).

UK

The 900 MTHM/yr THORP [Thermal Oxide Reprocessing Plant] was completed in 1992. It is located in Sellafield, England and operated by British Nuclear Fuels, PLC. The reported capital costs of THORP range from $5.3 billion to $7.5 billion in 1996 dollars. In 1993, British Nuclear Fuel Ltd. (BNFL) reported a cost of $5 billion ($1996), that excluded interest during construction (THORP was financed through up-front payments by its customers.) Of course, a decision to exclude these costs incorrectly omits the capital cost of returns forgone by the customers of THORP who provided the financing. Another 1993 estimate of THORP costs arrived at a figure of $4.8 billion in 1996 dollars (Nuclear Wastes: Technologies for Separations and Transmutation, 1996, pp. 418, 421).

An OECD/NEA report based on analysis of the THORP plant during construction estimated reprocessing costs in 1996 dollars of $720/kgHM for a 5% return on capital and $950/kgHM for a 10% per year return. Depending on assumptions, this indicates that the plant would have $7.6-$8.9 billion ($1996) in capital costs (Nuclear Wastes: Technologies for Separations and Transmutation, 1996, p. 421).

By 1996, the THORP plant had been operating for more than two years and had generated less than one metric ton of plutonium. Costs of the plant were officially listed as $4.4 billion and its operation was regarded as being responsible for 5000 jobs in the region. Government approval of THORP was based on the promise that it would make $820 million ($1996) profit in ten years. To do this, the plant should be reprocessing 700 metric tons of spent nuclear fuel a year toward its stated goal of 7000 tons in ten years. Thirty four companies in nine countries have reprocessing contracts with THORP, but THORP only managed to reprocess about 100 tons of spent fuel in 1995 and 1996 (Brown, 1996, p. 1).

Japan

The Rokkasho-mura Plant (800 MTHM/yr) is currently under construction. It is expected to begin operations in 2000. Japan Nuclear Fuel, Ltd. is constructing the plant with primarily French technology although there has also been both UK and German input. Despite the lessons of THORP and UP3, the constant dollar capital cost of Rokkashomura has not decreased relative to earlier plants. Early reported costs of the plant ranged from $6 to $7.3 billion in 1996 dollars and seismic problems significantly increased the design cost (Nuclear Wastes: Technologies for Separations and Transmutation, 1996, p. 416, 419). In January, 1996, Japan Nuclear Fuel announced that final construction costs of the Rokkasho-muro reprocessing plant are estimated to be $17 billion (Takagi, 1996, p. 7).

For any plant, the cost of production is based, in large part, on three costs related to the plant itself: the capital required to build the plant, the interest that must be paid on that capital, and the operating costs of the plant. Of these, the capital and interest expenses are fixed costs that must be paid whether the plant operates or not. A certain percentage of operating costs are also fixed, with the remainder being variable; i.e., changing with different production rates in the plant.

Reprocessing plant capital costs

Table 5.3 shows the reprocessing plant capital costs derived from a 1993 OECD report and based on actual data from THORP and from COGEMA's experience at UP3. These costs are based on aqueous (liquid) reprocessing costs since the alternative technologies of pyroprocessing, transmutation and the Advanced Liquid Metal Reactor (ALMR) are not sufficiently mature to allow cost estimation. This total estimated cost is higher than the reported cost of THORP, and this could be due to assuming construction at a more expensive site (Nuclear Wastes: Technologies for Separations and Transmutation, 1996, pp. 419-420). Table 5.4 shows capital costs listed in summarized by plant.

Table 5.3
Capital costs of the components of reprocessing plants

Cost Component	Capital Cost* (1996 $ millions)
Fuel Rcpt and Storage	220
Reprocessing Plant	4,960
High-Level Waste	
Vitrification	570
Interim Storage	130
Intermediate-Level Waste	
Encapsulation	650
Interim Storage	80
Site Preparation and Services	
Site Preparation	500
Site Services	23
Total Capital Cost	7,133

*Excludes Interest During Construction
Source: Organization for Economic Cooperation and Development (1993), Nuclear Energy Agency, *The Economics of the Nuclear Fuel Cycle*, OECD/NEA, Final Revised Draft, Paris.

Table 5.4
Capital costs by reprocessing plant - a summary of estimates

Reprocessing Studies	Base Yr For Cost Est.	Processing Throughput MTHM/yr[a]	Capital Cost 1996 $B	Unit Capital Cost $/kgHM/yr
Reprocessing Plants:				
THORP (UK-Sellafield)				
Wilkinson (1987)	1987	900	5.2	5,780
British National Fuel	1989	900	7.6	8,440
UP3 (France-LaHague)				
UP3 + UP2 Expansion	1985	800+400	6.5(UP3)	8,130
UP3 (Eng. & Const.)	1992	800	8.1	10,130
Rokkasho-mura				
Chang (1990)	1990	800	7.2	9,000
Uematsu (1992)	1989	800	5.8	7,250
Reprocessing Studies:				
Generic US (Gingold et al.)	1990	1,500	3.8	2,530
Pu Fuel (OECD/NEA, 1989)	1989	700	6.7-7.8	9,570-11,140
Nuclear Fuel Cycle:				
OECD/NEA, 1993	1991	900	6.0	6,670
Fuel Cycle Costs:				
BNFL, 1987	1987	600	3.5	5,830
ALMR Assessment:				
GE, 1991	1991	2,700	6.8	2,520[b]

[a]THORP plant capacity taken as 900 in all cases.
[b]This unit also provides fuel fabrication services so the capital cost of reprocessing would be slightly less than this figure.
Source: Committee on Separations Technology and Transmutation Systems, Board on Radioactive Waste Management, Commission on Geosciences, Environment, and Resources, National Research Council (1996), *Nuclear Wastes: Technologies for Separations and Transmutation*, Appendix J, National Academy Press, Washington, DC, p.423.

Interest charged during construction of reprocessing plants

Financing costs for reprocessing plants are large because of the lengthy construction times and the uncertainty and risk involved in reprocessing projects - particularly if the projects are privately owned. The UP3 and THORP plants both required 10 years to construct. The construction of Rokkashomura is scheduled to take 8 years. During construction, large amounts of capital are expended, and this capital must generally be borrowed because the plant is not ready to generate any revenues to offset expenditures. Thus, interest expenses are a significant added cost factor.

Interest costs were artificially depressed for the UP3 and THORP plants because the future customers of the plants provided the capital for construction before construction started. This removed the need to borrow construction money, but it also obscured the fact that there is an opportunity cost for financing in this manner that is the interest that could have been earned by the fund providers on an alternative investment. Viewed in this light, potential customers of the UP3 and THORP plants absorbed a surcharge equal to the anticipated return on the money they advanced for the project.

Table 5.5
Total interest costs during reprocessing plant construction

Owner	Constructed Cost	Interest During Construction*	Capital Cost*
Government	6,000	880	6,880
Utility	6,000	1,450	7,450
Private Venture	6,000	2,180	8,180

*All figures in millions of 1996 dollars
Note: Uncertainty is added to these figures by a lack of full information on the THORP and UP3 projects, exchange rate fluctuations over the life of the projects, labor productivity differences, and whether or not start-up costs were properly accounted for.
Sources: Organization for Economic Cooperation and Development (1993), Nuclear Energy Agency, *The Economics of the Nuclear Fuel Cycle*, OECD/NEA, Final Revised Draft, Paris, and Committee on Separations Technology and Transmutation Systems (1996), Board on Radioactive Waste Management, Commission on Geosciences, Environment, and Resources, National Research Council, *Nuclear Wastes: Technologies for Separations and Transmutation*, Appendix J, National Academy Press, Washington, DC, p .424.

The 1993 OECD/NEA cost study used a construction time of 11 years for reprocessing plants. Other facilities such as waste storage ponds, vitrification facilities and intermediate waste storage facilities were assumed to require 5-8 years to construct. Different capital costs were assumed for three types of facilities:

1. Government facilities where a 4% real interest cost was assumed.
2. Utility-owned facilities where a 6.4% weighted cost of capital was used.
3. Privately owned facilities where an 8% interest rate was used assuming project financing. The alternative to this method of financing - construction financing

- would accrue a prohibitively expensive 15.6% rate of interest (Nuclear Wastes: Technologies for Separations and Transmutation, 1996, pp. 422-424).

After assuming these interest rates, the OECD study applied a growth curve to allocate capital expenditures over the life of the construction project. The results, which give the total interest cost accumulated during construction, are shown in Table 5.5 (Nuclear Wastes: Technologies for Separations and Transmutation, 1996, pp. 423-424).

Operating costs

The third element of reprocessing costs is the operating cost of the plants. These costs are broken into the categories listed in Table 5.6.

Table 5.6
Reprocessing plant operating costs
(based on a 900 MTHM/yr annual throughput plant)

Cost Component	Operating Cost (Millions of 1996 $)
Fuel Receipt and Storage	24
Reprocessing Plant	311
High Level Waste	
Vitrification	47
Interim Storage	5
Intermediate-Level Waste	
Encapsulation	72
Interim Storage	2
Disposal	10
Low-Level Waste	
Disposal	24
Total Operating Cost	495

Source: Organization for Economic Cooperation and Development (1993), Nuclear Energy Agency, *The Economics of the Nuclear Fuel Cycle*, OECD/NEA, Final Revised Draft, Paris.

Profits from reprocessing

Given the costs presented in the previous sections, it is interesting to speculate on the levels of profit that could be realized by a reprocessor, assuming that operations proceeded as planned. Based on British Nuclear Fuel Ltd.'s (BNFL) own figures, income from THORP over the first ten years of its operation would amount to $12.2 billion ($1996) and costs would be $11.1 billion, yielding a profit of 1.1 billion even after accounting for the discounted $600 million cost of decommissioning. This leads to a profit of about $110 million per year on an investment of $5.7 billion, a rate of return of about 1.9%. Since its customers advanced most of the capital, the internal rate of return to BNFL is higher - "very approximately" 10% according to BNFL calculations. However, the 1.9% figure is a more accurate calculation because it is based on the total amounts of capital involved (Berkhout, 1993, p. 2). Further, as of 1996, the actual performance of the THORP plant has been so far below expectations that the 1.9% figure appears is very optimistic.

Conclusion: economics of reprocessing - the European experience

Between 1970 and 1980, the reprocessing prices experienced by Germany rose by a factor of ten - from DM 80-100/kgHM to DM 1800/kgHM - when measured in real terms. This led to concurrent rises in fuel and waste management costs so that by 1986 reprocessing costs made up about 40% of total fuel cycle costs. Over the last ten years, these costs have remained relatively stable contributing to an overall loss of competitiveness of nuclear power plants in Europe (Berkhout, 1996, pp. 38-39).

Recent studies by the OECD (1994) and the Energiewirtschaftlichen Institut (EWI)(1995) both show that a closed fuel cycle is less economical than an open fuel cycle with direct disposal. The OECD study gave a marginal advantage to the open cycle option while the EWI study showed a 25% advantage for the open cycle. In fact, if reprocessing costs are all attributed to the cost of MOX fuel, MOX is about six times as expensive as LEU fuel. This is calculated in 1996 dollars in the following manner:

The cost of a delivered LEU fuel assembly is $1000-$1500/kg LEU fuel. By comparison, MOX fuel fabrication and transportation costs are about $2000-$3000/kgMOX. Roughly 4 kg of spent fuel must be reprocessed to generate one kg of MOX. The cost of European reprocessing is now $1000/kgHM. Thus, it costs $4000 to reprocess enough material to generate one kg of MOX, creating a full cost range for MOX fuel of $6000-$7000 kg (Berkhout, 1996, pp. 40-41).

Table 5.7
General comparison of reactor fuel costs

Process	Cost
LEU:	
Uranium Conversion: (U_3O_8 to UF_6) Unit Cost	$9 ± $3/kgHM
Uranium Enrichment Unit Cost	$100 ± $25/kgSWU
Uranium Conversion (UF_6 to UO_2) Unit Cost	$8 ± $2/kgHM
LEU Fuel Fabrication Unit Cost	$225 ± $25/kgHM
Spent Fuel Disposal Cost	$460/kgHM
MOX:	
Fuel Reprocessing Unit Cost	$900 kg/HM*
MOX Fabrication Unit Cost	$1300-$1600/kgHM
Reprocessing High Level Waste Disposal Cost	$250-$1000/kgHM

* COGEMA and BNFL reportedly charged customers from $1400/kgHM to $1800/kgHM to cover capital costs of construction. Japanese costs were about $1600-$1700/kgHM in 1995.
Source: Cochran, Thomas B., Bowling, Miriam B., Paine, Christopher E. (1996), *The Cost of Russia's Civil Plutonium Separation Program*, Natural Resources Defense Council, Inc., New York, NY, June, p. 10-14.

In a 1996 study of Russia's civilian plutonium separation program, the Natural Resources Defense Council (NRDC) came to similar conclusions. Based on the 3000 megawatt thermal VVER-1000 reactor, the NRDC found the costs of LEU and MOX fuel fabrication were those shown in Table 5.7.

Table 5.8
VVER-1000 fresh LEU fuel costs ($/kgHM)

	Requirements (per kgHM fuel)	Unit Cost	Cost $/kgHM
Yellowcake (U_3O_8)	9.639 kg	$39 ± $13/kg	$276 ± $48
Conversion (U_3O_8 to UF_6)	8.219 kgHM	$9 ± $3/kgHM	$74 ± $25
Enrichment	7.460 kgSWU	$100 ± $25/kgSWU	$746 ± $186
Conversion (UF_6 to UO_2)	1 kgHM	$8 ± $2/kgHM	$8 ± $2
Fabrication	1 kgHM	$225 ± $25/kgHM	$225 ± $25
Total			$1330 ± $220

Source: Cochran, Thomas B., Bowling, Miriam B., Paine, Christopher E. (1996), *The Cost of Russia's Civil Plutonium Separation Program*, Natural Resources Defense Council, Inc., New York, NY, June, p. 14.

Based on these unit costs, the NRDC calculated the total costs of LEU fuel and MOX fuel for the VVER Reactor. As Tables 5.8 and 5.9 show, the use of LEU fuel has a cost advantage of four to five times over the use of MOX.

Table 5.9
VVER-1000 fresh MOX costs ($/kgHM)

	Requirements (per kgHM fuel)	Unit Cost	Cost $/kgHM
Reprocessing	4-5 kg spent fuel	$900/kgHM	$3600-$4500
Fabrication	1 kgHM	$1250-$1800/kgHM	$1250-$1800
Uranium Credit:			
U_3O_8	11-14 kg	$16-$20/kg	($176-$280)
Conversion	11-14 kg	$6-$9/kg	($66-$126)
SWU	4.4-5.5 kgSWU	$75-$100/kgSWU	($330-$410)
Total			$4000-$5700

Source: Cochran, Thomas B., Bowling, Miriam B., Paine, Christopher E. (1996), *The Cost of Russia's Civil Plutonium Separation Program*, Natural Resources Defense Council, Inc., New York, NY, June, p. 16.

These costs make it clear that MOX fuel has no economic benefit as a LWR fuel when compared to normal LEU fuel. However, there is also the question of the cost of directly placing spent fuel in final storage as opposed to reprocessing it for MOX fabrication. In this case, the economics of the situation are again unfavorable for any reprocessing-based activity. Table 5.10 provides German estimates of the cost of reprocessing spent fuel compared to direct disposal. It is interesting to note that in this comparison, reprocessing costs alone exceed the total costs for direct disposal.

All costs discussed in the above sections are based on experience or estimates derived from actual, operating plants or from the partially constructed Rokkashomura plant. As such, these costs reflect the best estimates of researchers and the only data currently available to accurately price the various options available to reprocessors. However, over the past twenty years a number of reprocessing cost studies have also been completed by various US researchers, many of whom are members of the nuclear industry or closely connected with companies related to the nuclear industry. Some of these US estimates are provided in the following section.

Table 5.10

German reprocessing costs compared with final storage costs

(1996 dollars per metric ton of spent fuel)

Operation	Reprocessing Cost	Final Storage Cost
Transport	$260,000	$260,000
Reprocessing and Waste Management	$4,544,000	$0
Plutonium and Uranium Utilization	$1,460,000	$0
Spent Fuel Storage and Conditioning	$0	$2,110,000
Definite Storage	$1,620,000	$1,950,000
Total	$7,884,000	$4,320,000

Source: Bauder, P., and Blaser, W. (1994), *Management of Operational (Including Spent Fuel and Reprocessing Waste from Nuclear Power Plants In Germany*, Fourth International Conference on Nuclear Fuel Reprocessing and Waste Management, RECOD '94, London, April 24-28.

US estimates of reprocessing costs

Reprocessing studies performed in the US in the 1990's, and the 1990 Advanced Liquid Metal Reactor (ALMR) reprocessing report all contain cost estimates that are well below (one-third to one-half) costs based on actual European and Japanese experience. A 1996 National Research Council (NRC) report concludes that the ALMR estimates "are far below the reported capital costs of actual plants. More important, they indicate an inverse economy of scale with plant throughput, which has not been observed or predicted in other studies." As a result, the NSC study "concludes that reported capital costs for actual contemporary plants currently provide the most reliable basis for estimating the cost of future plants. Estimated capital costs reported in recent US studies appear inexplicably low." (Nuclear Wastes: Technologies for Separations and Transmutation, 1996, p. 421)

Capital costs

The 1990 ALMR assessment calculated a capital cost of $7.4 billion for a 2,700 MTHM/yr reprocessing plant in 1996 dollars. This estimate included the cost of facilities for MOX fabrication that involved both aqueous PUREX technology and a newer TRUEX technology (Nuclear Wastes: Technologies for Separations and Transmutation, 1996, p. 421).

Another 1990 study of a generic US plant derived a capital cost that ranged between $3.3 billion in 1996 dollars for a government-owned plant and $3.6 billion in 1996 dollars for a private plant processing 1,500 MTHM/yr. These costs assumed a mature industry and the use of aqueous PUREX techniques (Nuclear Wastes: Technologies for Separations and Transmutation, 1996, pp. 419, 421).

Operating costs

Among the US studies, a 1991 paper by Gingold, et al, calculated an operating cost range of $206 million to $212 million in 1996 dollars for a 1,500 MTHM/yr plant (Gingold et al., 1991) and a 1991 report by GE claimed a cost of $300 million per year ($1996) for a 2,700 MTHM/yr plant. GE's cost included about $167 million per year for fuel fabrication. With this removed, the resulting operating cost of $130 million appears to be unreasonably low (Taylor et al., 1991). The $410 million ($1996) operating cost reported by the OECD is the only one based on actual operating experience and that has been subject to peer review (Nuclear Wastes: Technologies for Separations and Transmutation, 1996, p. 427).

As of 1996, the large US government-owned aqueous reprocessing facilities at the Savannah River Site were proposed to reprocess various kinds of spent fuel. Their cost in such a role should be lower given the fact that they are already constructed and, theoretically, have no capital costs to assign to the production process. A 1995 study by Westinghouse Savannah River estimated that these facilities could 'chemically stabilize' commercial nuclear spent fuels for about $9 billion. This figure included only $100 million in capital costs, and was primarily based on an operating cost that averaged $320 million per year in 1996 dollars for reprocessing 1000 MTHM (McWhorter et al., 1995, Sections 3.3, A1.0). This operating cost is only about 60% of the $495 million per 900 MTHM cost calculated for the THORP plant and 75% of the $410 million OECD estimate. While both THORP and the plants included in the OECD study had to amortize a portion of their capital costs in their annual operating expenses, they are also newer and theoretically, more efficient. Thus, the Westinghouse figures would appear to be somewhat optimistic.

Decommissioning costs

Few decommissioning costs are available for modern reprocessing facilities. However, these costs were provided for the THORP plant at Sellafield, England. In 1989, BNFL estimated these costs at $950 million ($1996). By 1992, decommissioning costs had risen to $1 billion ($1996) (Berkhout, 1993, p. 1).

Interest costs

As was the case in Europe, interest costs will vary depending on the mode of plant operation selected. Private and utility-owned plants will have less access to capital and will, as a result, experience higher interest rates than those facing government-owned projects. The NRC found the following issues were the principal determinants of the interest rates charged by prospective lenders for capital intensive reprocessing projects:

1. The immaturity of certain techniques (pyroprocessing and ALMR) and the need for successful demonstrations.
2. Likely growth in capital, operating and maintenance and decommissioning costs for pyroprocessing and ALMR technologies.
3. The difficulty of obtaining a government commitment for high cost transmutation technologies and in light of the potential health and safety effects.
4. Difficulty in attracting private capital due to perceived high technical, economic, and institutional risk of reprocessing and transmutation relative to alternative opportunities for investment (Nuclear Wastes: Technologies for Separations and Transmutation, 1996, p. 417).

For any investor to finance private ventures, the rate of return would have to be higher than that for current cogeneration projects because the risk is higher. At the present time, cogeneration projects experience the following financing statistics:

1. Debt to equity ratio: 80-85% debt, 15-20% equity because of low risk.
2. After-tax return on equity: 15-18%/yr (11-14% in constant dollars).
3. Nominal cost of long term debt: 10-11% (6-7% in constant dollars) or about 3% above the T-bill rate.
4. Financing supported by agreements with purchasers (Nuclear Wastes: Technologies for Separations and Transmutation, 1996, p. 428).

With these figures for reference, the interest rates charged to a private reprocessing venture would reflect high levels of risk arising from:

1. Unproved technology.
2. Likely public opposition.
3. High regulatory uncertainty.
4. Reluctance of finance community to participate.
5. Potential for adverse government policy.
6. The Barnwell experience (Nuclear Wastes: Technologies for Separations and Transmutation, 1996, p. 48).

Utility-owned facilities are also unlikely to succeed because utilities are generally afraid of nuclear projects because of their experiences with projects such as the Washington Public Power System (WPPS). There is ample evidence of this reluctance - no new nuclear plant has been started since 1978. As a result, the participation of utilities in any reprocessing plant appears highly unlikely (Nuclear Wastes: Technologies for Separations and Transmutation, 1996, pp. 429-430).

Government-owned facilities face a better interest rate environment because governments have a unique ability to get low financing rates. Further, the government is ultimately responsible for long-term disposal of high level waste and is the only institution that can overcome political issues. Hundreds of years of reprocessing may be required, and the federal government is the only institution that can guarantee such longevity (Nuclear Wastes: Technologies for Separations and Transmutation, 1996, p. 430).

As a result of these lower interest rates, government-run reprocessing plants can expect a much lower overall cost. When reprocessing is used in conjunction with light water reactor (LWR) technology a 1993 EPRI study of a 900 MTHM/yr, $7.8 billion ($1996) capital cost plant (including interest during construction), with an annual operation cost of $410 million, calculated the unit costs for each kilogram of heavy metal shown in Table 5.11.

Table 5.11
Unit costs for reprocessing plants

Plant Owner/Operator	Unit Cost $1996/kgHM
Government	880
Utility	1,450
Private	2,230
Source: Electric Power Research Institute (EPRI) (1993), *Technical Assessment Guide, Vol. 3, Electricity Supply*, EPRI, Palo Alto, California.	

Because of the obvious cost advantages inherent in government ownership of reprocessing facilities, the NRC Committee "concluded that if it were cost effective, government ownership of reprocessing and transmutation facilities would be the only realistic alternative." (Nuclear Wastes: Technologies for Separations and Transmutation, 1996, p. 444)

However, even with government ownership, there is no evidence that reprocessing leads to a cost effective method of reactor operation. A study of LWR fuel-cycle costs by the OECD in 1993 concluded that the levelized fuel cycle for once-through LWR use is about 14% less than a closed cycle using reprocessing, even when a credit was included for both the uranium and plutonium recovered through reprocessing (The Economics of the Nuclear Fuel Cycle, 1993). Partially in response to the poor economics of reprocessing and the closed fuel cycle, the US has chosen to terminate programs based on this technology. Budgeted US Department of Energy (DOE) amounts directly related to reprocessing and transmutation (S&T) technologies have followed the pattern shown in Table 5.12.

Table 5.12
Budgeted amounts for separation and transmutation technologies

Year	Amount
1992	About $75 million
1993	$133.5 million
1994	DOE begins phase-out of S&T programs
1995	$104.8 million for close out and termination costs.

Source: Committee on Separations Technology and Transmutation Systems (1996), Board on Radioactive Waste Management, Commission on Geosciences, Environment, and Resources, National Research Council, *Nuclear Wastes: Technologies for Separations and Transmutation*, National Academy Press, Washington, DC, p. 54.

Table 5.13
Summary costs: reprocessing price per kg heavy metal

Source	Avg. 10 yr Price 1996 Dollars
OECD/NEA Levelized Cost(1992)	$860/kgHM
DOE (GB and France, First Ten Years)	$1560/kgHM
DOE (GB and France, After First Ten Years)	$670-$780/kgHM
DOE (30 year wtd. avg.)	$970-1040/kgHM
Suzuki (Rokkasho-mura from 2000-2020)	$2000/kgHM
GOGEMA (First 10 Years - UP3)	$1400/kgHM
BNF Ltd. (Offer to German Firm from THORP)	$750/kgHM
Gingold et al. (Govt. Facility-1991)	$275/kgHM
Gingold et al (Private Facility-1991)	$560/kgHM
GE (ALMR Study Extrapolation)	$285/kgHM
National Research Council (Levalized Cost, 30 Year plant life - Govt. Facility)	$910/kgHM

Sources: Organization for Economic Cooperation and Development (1993), Nuclear Energy Agency, *The Economics of the Nuclear Fuel Cycle*, OECD/NEA, Final Revised Draft, Paris, and Committee on Separations Technology and Transmutation Systems (1996), Board on Radioactive Waste Management, Commission on Geosciences, Environment, and Resources, National Research Council, *Nuclear Wastes: Technologies for Separations and Transmutation*, Appendix J, National Academy Press, Washington, DC, p. 433.

Summary: a comparison of published reprocessing prices

When one adds the annual principal and interest payments for capital to the annual operating costs for a plant and then divides this sum by the number of kg of heavy metal processed in a year, the result is the cost per kilogram to reprocess spent fuel. Since interest and operating costs will vary over the first few years of operation of any plant, a more conservative cost per kilogram may be generated by taking an average of the annual costs over a ten-year period. As Table 5.13 shows, reprocessing prices calculated in this manner generally range between $750 and $2000/kgHM with the lower prices reflecting either the lower cost of capital associated with the THORP and UP3 plants or estimates made by pro-nuclear sources in the US.

According to the National Research Council, the cost comparisons shown in Table 5.7 call into "question the validity of all the recent US estimates for the cost of reprocessing LWR spent fuel. The estimates for aqueous [reprocessing] are far below the costs inferred from the European and Japanese benchmarks." (Nuclear Wastes: Technologies for Separations and Transmutation, 1996, p. 434) In fact, based on the more realistic European-based costs in Table 5.9, it would cost from $50 billion to $130 billion to reprocess the 62,000 metric tons of spent fuel destined for Yucca Mountain at a US reprocessing facility (Nuclear Wastes: Technologies for Separations and Transmutation, 1996, p. 78).

A comparison of reprocessing costs with the costs of burning nuclear reactor fuel in a once-through, direct disposal scenario shows that reprocessing makes the use of nuclear reactors much more expensive. The OECD/NEA study calculated levelized LWR fuel cycle costs for reprocessing and for direct disposal. The levelized cost of reprocessing exceeded that for direct disposal by 12%. Using the OECD figures, the breakeven reprocessing cost in 1996 dollars would be $420/kgHM with a 5% discount rate and $355/kgHM at a 10% discount rate (The Economics of the Nuclear Fuel Cycle, 1993, and Nuclear Wastes: Technologies for Separations and Transmutation, 1996, pp. 435-436).

One major issue in these costing comparisons is how sensitive the costs are to changes in changes in assumed plant capital and operating costs, financing rates, financial structure, and currency exchange rates. For example, using standardized plans for a series of ALMR plants [an approach similar to the European method of building plants] would reduce the capital cost by $500 million to $1 billion (Nuclear Wastes: Technologies for Separations and Transmutation, 1996, p. 436).

Thus, if the cost of reprocessing is too high, it would be cheaper to simply start first generation ALMR's with virgin uranium enriched to about 30-40% U-235. Under certain pricing scenarios, this might be a viable approach for producing electricity if the enriched uranium is not too expensive, but it is not a viable means of transmuting the weapon-grade materials present in spent fuel. Assuming the $16/LB cost for U-3O8 and separative work unit cost of $95/kg SWU commonly available in 1996, the threshold reprocessing cost would have to be $315/kgHM or less to make the use of reprocessed fuels viable. If reprocessing exceeds this figure, and Table 5.7 shows it is likely to be double or triple this amount, enriched uranium is cheaper (Nuclear Wastes: Technologies for Separations and Transmutation, 1996, p. 440). And if enriched uranium is cheaper, it would be more cost effective to simply use LWRs.

Based on this type of cost comparison, the target cost for pyroprocessing must be about $315/kgHM or less. To accomplish this, the Argon National Laboratory Integral Fast Reactor concept would have to reduce - *by a factor of six to seven times* - the cost of large scale aqueous reprocessing. The disparity in cost would be even greater if the TRUEX process was added to obtain high recoveries for transmutation. Studies done at Oak Ridge show that even at $315/kgHM it would be cheaper to build and operate a once-through LWR instead of ALMRs (Nuclear Wastes: Technologies for Separations and Transmutation, 1996, p. 440).

Thus, there is no credible evidence that use of reprocessing in any form is likely to produce an economical alternative to once through LWR use. As the National Research Council notes,

> The limited experience with pyrochemical reprocessing of LWR fuel as proposed by the ALMR program offers no compelling argument that current cost estimates will not increase as detailed designs emerge or that such reprocessing would be less costly than conventional aqueous reprocessing of LWR fuel, or even less costly than a PUREX facility augmented with the aqueous TRUEX process....(Nuclear Wastes: Technologies for Separations and Transmutation, 1996, p. 443)

Further, there is no indication that any of these costing studies, unfavorable as they were, accurately accounted for the full costs of the wastes produced during reprocessing. The generation of such wastes, particularly from aqueous reprocessing, is something with which the US has considerable experience. The following section deals with these costs.

Waste generation and US defense nuclear reprocessing

Large scale reprocessing of military spent fuel to extract plutonium for bombs continued for a considerable period of time at Hanford, Washington, and at the Savannah River Site (SRS) in South Carolina. Additional reprocessing to recover uranium for military purposes also occurred at both these sites and at the Idaho Chemical Reprocessing Plant at the Idaho National Engineering Laboratory. All of these sites were closed as of 1996 with the exception of the H Canyon site at SRS. Costs to operate the H Canyon were about $170 million per year in 1996. If the facility were placed in standby, the annual costs of maintenance would be about $165 million per year. The DOE has recommended consolidating operations at the F Canyon instead, claiming that this would lead to a ten-year savings of $162 million (Ford, 1996). However, no credible costing evidence of this claim has been given and no hard evidence of the need to use either canyon has been provided, although DOE did recommend the continued reprocessing of "small quantities [of research reactor fuel types that] could be accommodated easily at SRS while the canyons are still available for other uses." (Technical Strategy for the Treatment, Packaging, and Disposal of Aluminum-Based Spent Nuclear Fuel, 1996, p. 4)

Table 5.14
Spent fuel inventories at various storage sites

Location	1995 Amount	1995 Percent	2035 Amount	2035 Percent
Hanford	2133	80.60	2103	76
INEL	261	9.80	426	16
Savannah River Site	206	7.80	213	8
Oak Ridge Reservation	1	.03	0	0
Other DOE Facilities	27	1.00	0	0
Universities	2	.08	0	0
Other	16	.60	0	0
Total	2646	100	2741	100

Source: US Department of Energy (1995), *Record of Decision, Department Of Energy Programmatic Spent Nuclear Fuel Management and Idaho National Engineering Laboratory Environmental Restoration and Waste Management Programs*, DOE Office of Environmental Management and Idaho Operations Office, May 30, pp.8, 13.

US DOE has reprocessed more than 100,000 metric tons of spent fuel at a very high cost to satisfy national defense requirements (Record of Decision, Department Of Energy Programmatic Spent Nuclear Fuel Management and Idaho National Engineering Laboratory Environmental Restoration and Waste Management Programs, 1995, p. 7). In the process, little thought was given to the amount of waste material produced by the reprocessing. For example, a reprocessing facility at Hanford, Washington, produced 55,000 gallons of liquid radioactive waste and 340 gallons of liquid high-level waste for every kilogram of plutonium produced (Nuclear Reactions, 1996).

Generation of sizable amounts of waste through reprocessing could continue under current plans. DOE has an inventory of 2700 metric tons of spent fuel, much of which was not meant to be stored for a long period. This spent fuel is corroding and releasing radioactive material into the cooling water in which it is stored. Reprocessing the solid spent fuel at the Savannah River Site alone would result in about 3 million gallons of high-level liquid waste (Sachs, 1996, pp. S-1, S-4, 2).

In a current effort by the DOE to restart reprocessing, reprocessing is funded in DOE's Fiscal Year 1997 budget request at about $300 million at the Savannah River

Site in South Carolina and $50 million at Argonne National Laboratory-West at the Idaho National Engineering Laboratory (INEL) (Nuclear Reactions, 1996). At the Savannah River site, this 'modern' reprocessing is likely to generate almost as much waste as previous US programs did. Table 5.15 shows the amounts of waste generated over 10 years of operations under alternatives for stabilizing F-Canyon plutonium solutions by reprocessing and by vitrification at the Defense Waste Processing Facility (DWPF).

Table 5.15
Waste generation under various alternatives
at the Savannah River site - in cubic meters

Type of Waste	Process to Metal	Process to Oxide	Vitrification at DWPF	Vitrification at F-Canyon
Saltstone	5689	6461	4813	5592
Transuranic	60	175	4	147
Low-Level	11,907	14,371	10,174	13,820

Source: Department of Energy (1994), *Final F-Canyon Plutonium Solutions Environmental Impact Statement*, DOE/SRS, Aiken, SC, December, pp. 4-28.

Alternative reprocessing technologies for separating plutonium only - proliferation concerns

Reprocessing old spent fuel to recover plutonium differs in two ways from reprocessing fresh fuel to recover both plutonium and uranium or uranium only. First, the radioactivity of old spent fuel is lower, reducing shielding requirements and making handling easier. Second, if plutonium recovery is the only issue and recovery of uranium is not required, simplified chemical processes are involved (Peterson, 1995, p, 435). However, since uranium can be sold to help finance reprocessing, the decision not to recover uranium lowers the potential revenues from reprocessing.

In addition, the capital cost of a reprocessing facility used solely for separating plutonium for weapons can be lower than that of a power production-related [uranium recovery] reprocessing facility because of simpler design - particularly if health and safety standards are reduced. The Office of Technology Assessment estimated that a small reprocessing facility could be built for as little as $62 million (1996 dollars) (Nuclear Proliferation and Safeguards, 1977, pp. 140, 177, 180).

When dealing solely with plutonium production for weapons, there are other inexpensive ways to proceed. For example, a small 25 MW graphite-moderated reactor could produce about 5 kg of Pu per year at a 60% operating rate. A dedicated 400 MW reactor, capable of producing 100 kg/year of plutonium with an accompanying reprocessing facility, would require 50-75 engineers, 150-200 skilled technicians, 5 to 7 years from start of design to first output, and a capital outlay of $435-$870 million (1996 dollars) (Nuclear Proliferation and Safeguards, 1977, pp. 140, 177, 180).

Notes

1   The spent fuel standard proposes to make plutonium as difficult to retrieve as it would be if it were in the form in which it exists in nuclear reactor fuel that has been irradiated (used) to the extent it can no longer effectively sustain a chain reaction and thus, has been removed from the reactor for disposal. This irradiated fuel contains fission products, uranium, and transuranic isotopes.

2   The term 'heavy metal' is usually employed to describe the spent fuel submitted to the plant for reprocessing. Reprocessing plants are rated in terms of their capacity in heavy metal, with the acronym 'MTHM/yr' denoting the metric tons of heavy metal per year capable of being processed by the plant.

References

Berkhout, Franz (1993), *Fuel Reprocessing At THORP: Profitability and Public Liabilities*, Center for Energy and Environmental Studies, Princeton University, Princeton, NJ, pp. 1, 2, 12.

Berkhout, Franz (1996), 'The Rationale and Economics of Reprocessing', in *Selected Papers From Global '95 Concerning Reprocessing*, W.G. Sutcliffe, Ed., UCRL-ID-124105, Lawrence Livermore National Laboratory, California, pp. 38-39.

Brown, Paul (1996), 'Production Crisis Hits THORP Nuclear Plant', *Manchester Guardian*, August 27, p. 1.

Closing the Circle on the Spitting of the Atom (1996), U. S. Department of Energy, Office of Environmental Management, January, pp. 30, 31.

Draft Environmental Impact Statement on a Proposed Nuclear Weapons Nonproliferation Policy Concerning Foreign Research Reactor Spent Nuclear Fuel (1995), DOE Office of Environmental Management, US Department Of Energy, Washington, DC, pp. 2-16.

Ford, John (1996), *DOE Briefing*, Office of Nuclear Materials and Facility Stabilization, DOE.

Gingold, J.E., Kupp, R.W., Schaeffer, D., Klein, R.L. (1991), *The Cost of Reprocessing Irradiated Fuel From Light Water Reactors: An Independent Assessment*, NP-7264, EPRI, Palo Alto, California.

Management and Disposition of Excess Weapons Plutonium (1994), Committee on International Security and Arms Control, National Academy of Sciences, National Academy Press, Washington, DC, p. 2.

McWhorter, D.L., Geddes, R.L., Jackson, W.N., Bugher, W.C. (1995), *Chemical Stabilization of Defense Related and Commercial Spent Fuel at the Savannah River Site*, Document No. NMP-PLS-950239, Westinghouse Savannah River Company, Sections 3.3, A1.0.

Nuclear Proliferation and Safeguards (1977), Office Of Technology Assessment, (Library of Congress Catalog No. 77-600024), pp. 177, 180, 140.

'Nuclear Reactions' (1996), *Washington Post Magazine*, May 5.

Nuclear Wastes: Technologies for Separations and Transmutation(1996), Committee on Separations Technology and Transmutation Systems, Board on Radioactive Waste Management, Commission on Geosciences, Environment, and Resources, National Research Council, Appendix J, National Academy Press, Washington, DC, pp. 48, 78, 335, 412-444.

Peterson, Per F. (1995), 'Long-Term Retrievability and Safeguards for Immobilized Weapons Plutonium in Geological Storage', *Final Proceedings: US Department of Energy Plutonium Stabilization and Immobilization Workshop*, p. 435.

Protection and Management of Plutonium (1995), American Nuclear Society Special Report, August, pp. 1, 2, 10.

Record of Decision, Department Of Energy Programmatic Spent Nuclear Fuel Management and Idaho National Engineering Laboratory Environmental Restoration and Waste Management Programs (1995), DOE Office of Environmental Management and Idaho Operations Office, US Department of Energy, p. 7.

Sachs, Noah (1996), *Risky Relapse into Reprocessing*, Institute for Energy and Environmental Research, pp. 2, 25-31, S-1, S-4, S-5.

Seaborg, Glenn T. (1995), 'Preface', *Protection and Management of Plutonium*, American Nuclear Society Special Report.

Takagi, Jinzaburo (1996), *Japan's Plutonium Program and Its Problems*, Third International Radioecologica Conference on The Fate of Spent Nuclear Fuel and Reality, Krasnoyarsk, Russia, p. 7.

Taylor, I.N., Thompson, M.L., Wadekamper, D.C. (1991), *Fuel Cycle Assessment-1991*, GEFR-00897, General Electric, San Jose, California.

Technical Strategy for the Treatment, Packaging, and Disposal of Aluminum-Based Spent Nuclear Fuel (1996), Office of Spent Fuel Management, Department of Energy, p. 4.

The Economics of the Nuclear Fuel Cycle (1993), Organization for Economic Cooperation and Development, Nuclear Energy Agency, OECD/NEA, Final Revised Draft, Paris.

6 Disposition through MOX burning: fabrication and operations issues

Introduction

Mixed oxide fuel (MOX) is nuclear reactor fuel fabricated from plutonium. The first tests of MOX were conducted in Belgium in 1963. By 1996 about 400 metric tons of MOX had been burned in commercial reactors, producing 100 billion kWh of electricity. In Europe, 18 pressurized water reactors and boiling water reactors have at one time or another been loaded with MOX (Rougeau, 1996). Historically, the use of MOX was never meant to provide a rationale for the production of plutonium. MOX was viewed, instead, as a temporary measure to bridge the period between the use of currently available plutonium stocks and the time when fast breeder reactors became operational. However, since almost all fast breeder reactors have been canceled, MOX has taken on a life of its own (Schneider, 1996, p. 4).

All nuclear fuel containing plutonium is called MOX, including fuel manufactured for Fast Breeder Reactors (FBRs), but proposals to use MOX fuel usually involve light water reactors (LWRs). The plutonium content of MOX can vary substantially depending on whether LWRs or FBRs are used. For example, the amount of plutonium in FBR fuels varies between 20 and 35%, while the amount in the MOX for LWRs varies from 4 to 7% (Kueppers and Sailer, 1994, Ayukawa, 9/23/1996).

The use of plutonium in civilian reactors is not now widespread because plutonium creates no economic benefits and its use is accompanied by a large proliferation risk. Thermal cycle use (reprocessing plutonium and uranium from spent fuel and using MOX fuel in thermal nuclear power plants) and plutonium-fueled fast reactors will not be competitive with the use of low enriched uranium (LEU) in LWRs until the price of uranium-bearing yellowcake reaches $110/LB in 1996 dollars - a price Chow and Solomon estimate will not occur for 50 years (1993, pp. xvi, xvii). As a result, one cannot discuss the fabrication and use of MOX without acknowledging that this use is normally predicated on the replacement of some amount of uranium fuel in LWRs and it usually compared with the cost of using LEU fuel in similar applications.

74

Uranium production fell and spot prices declined for nine consecutive years between 1985 and 1994. Production increased 4.4% in 1995, but this level of production still represented little more than 50% of annual reactor requirements. Uranium spot market prices rose enough in 1995 to stimulate uranium mining activity (Ashton, 1996). However, in October, 1996, the spot price of uranium was still only $16.30/lb. and prices were falling again due to the presence of new uranium sellers (Nuclear Fuel, 1996, p. 2). New supplies of uranium seem likely to keep spot prices depressed.

Australia, Canada, Niger and the United States led the world in uranium production in 1995. Niger's production is not expected to change greatly in the near future, but the other three countries are planning to expand uranium production in the coming years, each for different reasons. In Australia the main stimulus to increased production is the end of the former government's Three Mines Policy. This allows Australia's uranium mines to seek approval for new projects. As a result, for the first time since 1983-84, Australia's total uranium production exceeded 4240 tons in 1995-96 and should increase to about 8000 tons (Ashton, 1996).

In Canada, production expansion is part of a long-term plan to maintain supplies. In 1995 Canada's uranium production reached 10,515 tons, its highest level since 1989, and an 8.5% increase over the 1994 figure. Canadian development plans are part of a long-term strategy to extend capacity and maintain Canada's position as the world's premier supplier of uranium. Production capacity is anticipated to continue to rise beyond the year 2000 (Nuclear Fuel, 1996, p. 2).

Planned increases in US uranium production are almost exclusively price-driven. In 1994, the United States produced only 1400 tons of uranium. This increased to 2360 tons in 1995 in a turnaround that followed years in which the industry in the United States had been in steep decline.

The uranium supply/demand gap closed in 1996 and is likely to narrow further over the next few years (Nuclear Fuel, 1996, p. 2). However, even with the recent increases in uranium production, some forecasters feel that world demand will continue to outstrip world supply. On September 6, 1996, the London-based Uranium Institute's market report on future uranium supply and demand "concluded that only with the combination of the lowest requirements scenario and the highest supply scenario will uranium production be sufficient to meet demand, and then only from 2002 onwards. Otherwise demand may exceed planned supply by as much as 15,000 tons of uranium per year." (Ashton, 1996) This report predicts that in the year 2000 the lowest estimate of requirements would be about 65,600 tons of uranium, while the highest projected supply figure would be 61,400 tons. By 2010, the lowest predicted requirements are 66,900 tons, with supply ranging from 58,300 to 70,100 tons (Nuclear Fuel, 1996, p. 2).

If a demand gap does indeed exist, one obvious source of uranium to fill it would be the weapon-grade HEU from excess nuclear warheads. A study by the ANSALDO-ENEA-ENEL working group forecasts a 400 GWe nuclear capacity in

2005. 85% of this capacity would be in LWRs. The annual requirement for uranium oxide U_3O_8 will rise from 50,000 tons in 1990 to 70,000 tons in 2005. Dismantling one third of all warheads in the next 15 years will supply the equivalent of 12,000 tons of uranium oxide enriched at 3.3%. This would be 11% of all uranium needed to fuel all reactors during this period. Burning all warheads would generate one year of the earth's requirement for electricity - 10 trillion kWh. To generate 6.5 billion kWh with coal fired power plants of 1000 MWe would create 6 million tons of CO_2, 31,600 tons of SO_2, 2,500 tons of CO, 18,300 tons of NOX, and 2,400 tons of particulate and powder per year. For oil fired plants, 4.41 million tons of CO_2, from 20,000 to 80,000 tons of SO_2, 2,200 tons of CO, 8000 tons of NOX, and 130 tons of powders and particulates would be created (Cumo, 1992, pp. 95-97).

MOX use and subsidies

Utility interest in MOX

Nuclear reactors are operated to generate power, and power generation is a commercial enterprise. A commercial enterprise can only operate if it makes profits and to do this, operating costs must be kept as low as possible. Thus, a central issue concerning MOX use is whether it is cheaper or more expensive to use than the conventional low enriched uranium (LEU) fuel currently used in reactors. This question has been extensively researched and the answer is unambiguous - MOX is more expensive. In fact, in the early 1990's, the costs for MOX were five to six times those for LEU-based reactor fuel (Berkhout, 1993, p. 6). Even DOE's own *Report For Surplus Weapons-Usable Plutonium Disposition* states that "in no case can MOX fuel complete [sic] economically with LEU fuel." (1996, p. 4-6)

Given this unfavorable cost difference, no commercial reactor operator would willingly accept MOX fuel unless that operator was subsidized by the US government at a level sufficient to compensate for the losses caused by choosing MOX. Such a subsidy is called an 'incentive fee' by the Department of Energy, and the size of this subsidy will be the deciding factor when a commercial operator is given the option of using MOX.

In December, 1995, DOE "issued a request for expressions of interest for tritium production that also solicited interest in regarding the future potential use of mixed oxide fuel from surplus weapons plutonium either coincident with or separate from tritium production." (Technical Summary Report For Surplus Weapons-Usable Plutonium Disposition, 1996, p. 4-7) Table 6.1 shows the expressions of interest this request elicited in 1996.

According to Greenpeace International, Florida Power and Houston Lighting and Power later decided not to take part in either part of this program, and Arizona Public Service responded to the DOE's inquiry only "to obtain additional information on these programs," and "has not volunteered to produce tritium for

DOE." (Arizona Public Service Company, 1996) Commonwealth Edison and Duke Power said they have aligned with COGEMA and BNFL to study burning MOX in their reactors (Nucleonics Week, 1996).

Table 6.1
Utility and private company interest
in MOX burning and/or tritium production

Utility	Tritium Production	MOX Burning
AZ Public Service Co.	yes	yes
Centerior Energy of OH	yes	yes
Duke Power/Commonwealth Edison of MO	no	yes
Energy Operations Inc. of LA	no	yes
FL Power and Light Co.	yes	yes
GA Power Co.	yes	yes
Houston Lighting & Power	yes	no
IES Utilities, Inc. IA	no	yes
IL Power Co.	yes	no
Niagara Mohawk Power Co. of NY	yes	yes
NC Municipal Power Agency		
#1/Piedmont Municipal Power Agency	yes	yes
PECO Energy Co. PA	no	yes
SC Electric and Gas	yes	no
Southern Nuclear Operating Co. of AL	no	yes
Tennessee Valley Authority	yes	yes
VA Power	yes	yes
WI Public Service Co.	yes	yes
WA Public Power Supply System	yes	yes
Private companies		
Westinghouse	yes	yes
ABB-Combustion Engineering	yes	yes
Utility Resource Associates	yes	yes
AECL Technologies/Team CANDU	no	yes
COGEMA	no	yes
BNFL	no	yes
Belgonucleaire	no	yes
Lockheed Martin INEL	no	yes

Source: DOE (1996), March 28, and Greenpeace International Press Release (1996), March 29.

The responses in Table 6.1 were predicated on the prospect of both free MOX fuel and a possible subsidy from the US government (Numark, 1996, p. 6). In conjunction with this solicitation of interest, DOE compiled a detailed list of the estimated charges it would be willing to cover in return for a utility's agreement to use MOX fuel. These costs were estimated to be approximately $825 million per reactor through 2024, with the greatest cost being the waiver of the utilities' contribution to the Nuclear Waste Fund, a sum equal to $310 million per reactor (National Conference of State Legislatures, 1996).

The DOE has stated that it regards these responses as evidence of "sufficient commercial interest in use of existing or partially completed light water reactors for plutonium disposition..." (Technical Summary Report For Surplus Weapons-Usable Plutonium Disposition, 1996, p. 4-7) A more even-handed view of the responses to the DOE offer would note that shifting all MOX fuel fabrication costs to the US government - i.e., providing free MOX fuel - did not provide a sufficient economic incentive to make a power producer switch from LEU fuel. Instead, an incentive fee (subsidy) of indeterminate size was still required to make MOX competitive with LEU. This means that reactor owners were not expressing a 'commercial interest' in the normal sense of private enterprise - they were, instead, offering to allow the US government to use their facilities for a fee. With this in mind, it is important to consider exactly how large the subsidy would have to be before a utility would allow the US government to use its plant.

The need to subsidize MOX use

Because of the negative connotations associated with government subsidies of ostensibly private commercial enterprises, the DOE has chosen to call the subsidies it proposes for MOX use "incentive fees." To be acceptable to the power producer, a subsidy for MOX-burning would have to make the power producer competitive with alternative sources of energy. Thus, a correctly-sized subsidy would allow rate-payers to pay the same charge for electricity whether they purchased it from the nuclear power producer or from a coal-fired power producer who was providing competitive prices in a free market environment.

However, when nuclear power is involved, it is not clear how this concept would be applied. Industry analysts have always compared the costs of nuclear power generation with the costs of all competing sources of energy. As deregulation of the power industry continues, these cost comparisons have become more meaningful because a deregulated industry will have greater difficulty 'covering' the costs of one or two marginal power generators.

However, the DOE has avoided making comparisons between MOX use and any other competing source of power except other light water reactors. When the rationale for the use of one form of energy depends solely on the relative price level of just one of a number of competing energy sources as it does in this case, one encounters the 'energy value trap' discussed earlier. For example, claiming that MOX becomes an economically viable source of energy when the price of uranium dramatically increases misses the economic point that MOX becomes economically viable only when it can successfully compete with all other energy sources. If other sources of energy have stable prices, the fact that uranium's price is increasing is not relevant to the economic viability of MOX - it only indicates that uranium is becoming less competitive.

For example, in 1996 the DOE's *Report For Surplus Weapons-Usable Plutonium Disposition* recorded operating costs of MOX-burning facilities only as the net additional costs of LWRs burning MOX fuel. None of the significant costs to operate a LEU-burning light water reactor were included, and DOE declined to include "estimates of incentive fees, if any, that might be paid to utilities for MOX irradiation services...in addition to the expected reimbursable costs that would be incurred by the utilities..." (1996, p. 4-3) If one assumes that LEU-powered reactors were competitive with other energy sources at the time this report was written, these net additional costs can be viewed as an estimate of the levels of subsidy that would have to be provided to the operating costs of a MOX user to encourage plutonium burning for power production. However, subsidies to operations are only one of a number of subsidies that the DOE would have to deliver to MOX users. If one assumes the DOE figures are accurate, these subsidies can be calculated from Table 6.2, which shows the costs of various MOX fuel options as calculated by the DOE.

Table 6.2

1996 DOE-generated costs for MOX use - millions of 1996 dollars

Existing Reactor Alternatives	Facility	Investment	Operating[3]	Fuel Displcmt Credit[5]	Life Cycle Cost If Financing Is By Govt.	Private
LWR's	Front-end[4]	340	1050	0	1390	1390
	MOX Fab[4]	410	1130	-1390	150	750
	Reactor	230	150	0	380	380
	Total	980	2330	-1390	1920	2520
LWR's Greenfield Facilities[1]	Front-end[4]	1050	2590	-2010	1630	1630
	Reactor	330	130	0	460	460
	Total	1380	2720	-2010	2090	2090
CANDU	Front-end[4]	320	1090	0	1410	1410
	MOX Fab[4]	450	1430	-320	1560	2160
	Reactor	100	40	0	140	140
	Total	870	2560	-320	3110	3710

Partially Complete & Evolutionary Reactors	Facility	Investment	Operating[3]	Revenues[2]	Life Cycle Cost With Revenues	Without Revenues
Partially Complete LWR's	Front-end[4]	320	1090	0	1410	1410
	MOX Fab[4]	350	1120	0	1470	1470
	Reactor	2380	2400	-7890	-3110	4780
	Total	3050	4610	-7890	-230	7660
Evolutionary LWR's	Front-end[4]	320	1090	0	1410	1410
	MOX Fab[4]	350	710	0	1060	1060
	Reactor	6210	2980	-7150	2040	9190
	Total	6880	4780	-7150	4510	11660

[1]Greenfield front-end and MOX fuel fabrication facilities are co-located and their costs are combined.

[2]These revenues accrue to the owners of the plant, not to the government. Thus, their impact on the plutonium disposition mission costs is unknown.

[3]Operating costs recorded in this table are only the net additional costs for MOX operations. None of the costs to operate a LEU-burning plant are included.

[4]Front-end costs + MOX Fab costs = cost of plutonium processing and MOX fabrication.

[5]The cost of LEU fuel displaced by MOX fuel.

Sources: Author's Calculations and the *Technical Summary Report For Surplus Weapons-Usable Plutonium Disposition* (1996), DOE/MD-0003, Office of Fissile Materials Disposition, US Department of Energy, Washington, DC, pp. 4-1 to 4-8 and *Revision 1 to Technical Summary Report For Surplus Weapons-Usable Plutonium Disposition* (1996), October 10.

There is ample evidence that nuclear plants are marginally competitive with other forms of power production in the United States (and in some cases, they may already be non-competitive). In the most recent example, Northeast Utilities announced in October, 1996 it was shutting down the Connecticut Yankee nuclear power plant because the costs and benefits of running the plant for another 10 years, "[do] not seem favorable." A worst case scenario showed an "economic disadvantage" of about $100 million if the plant continued operations. The lower market price of fossil fuel power generation was responsible for some of this 'disadvantage' and the high cost of maintaining the plant accounted for the rest (Kerber, 1996, p. A6). Canada's CANDU reactors are also non-competitive. They produce more expensive electricity than Canadian fossil fuel plants and they, too, will require expensive maintenance to avoid an early shutdown. In fact, one of these reactors (Bruce 2) suffered damage to its steam generators and was mothballed in 1995 to avoid costly repairs (Silver, 1996, p. 8). 'Normal' steam generator replacement for the CANDU is very straight-forward and has been done at about 30 plants world-wide. However, even this operation usually carries a total cost of $100-$200 million (Fuoto, 1996).

In addition to their general non-competitiveness, a number of other factors make nuclear power in the US only marginally competitive or non-competitive in situations where European nuclear power claims to be competitive. These factors are unlikely to change in the future, and many of them have been complicated by the introduction of new, unproved technologies. These factors arise from:

1. Non Standard Reactor designs. Past subsidies to the nuclear industry in the US have gone almost entirely to research and development. The predictable result was a proliferation of reactor designs. The US currently suffers from very little standardization of design, a factor that raises operating, repair, and refueling costs. Europe, on the other hand, has much more standardization by country.
2. Different rates or costs for competitive fuels. Competitive means of power generation are generally more available and cheaper in the US than they are in Europe. This makes it more difficult for nuclear reactors to compete in US power markets.
3. European tendencies toward government-owned power companies. US facilities, on the other hand, are often run by private power producers, a fact that raises the cost of capital significantly and requires that costs be treated in a more transparent fashion.
4. European financing advantages. European sites have had up-front financing for a significant part of their construction costs. This financing, given by future customers of European reprocessing and MOX facilities, significantly lowered the recorded cost of capital (although, not the actual cost of capital) of the European sites.

5.   Waste management differences. While the original reprocessing contracts left the waste generated by reprocessing and MOX fabrication with the country in which reprocessing occurred, waste generated by recent contracts with European plants is sent back to the country for whom the reprocessing was done. Shipping waste back to its country of origin relieves European reprocessors/MOX fuel fabricators of the costs of waste disposal. US operations must account for the cost of waste disposal through a fee that is added to the costs of power sold by the facilities.
6.   Citizen involvement and regulatory oversight. In the US, these factors have undoubtedly raised costs in the short term. This has adversely affected the competitiveness of US reactors by making the specific costs of operation more apparent than they are in Europe. However, these same factors should result in lower long-term costs of nuclear power in the US.

When nuclear power plants are already having trouble remaining competitive for reasons such as these, any additional costs (such as maintenance, additional security, or modifications) further degrades the ability of these plants to compete with other commercial power producers. For this reason, either the subsidy required to induce nuclear plant owners to use MOX must cover all additional costs of MOX use or the additional costs must be shifted to those who purchase the power. As a result, the subsidies to nuclear plants for using MOX must occur in three areas, investment, operations and capital costs.

Subsidies to investment in facilities

MOX use requires new production facilities and increased safeguards in existing facilities because of the sensitive nature of plutonium. None of these costs would be incurred with standard LWR operations. Thus, the life cycle cost of federal subsidies in this area can be calculated in Table 6.2 to be:

Existing Reactors: $870 million to $1.38 billion
Evolutionary and Partially Completed Reactors: $3.05 billion to $6.88 billion

Subsidies to operations of facilities

The DOE claims that all operations costs in Table 6.2 are in addition to the cost of using LWRs with LEU. As a result, all of these costs would have to be subsidized to allow reactor operators to maintain their original competitiveness when using MOX fuel.

Existing Reactors: $2.33 billion to $2.72 billion
Evolutionary and Partially Completed Reactors: $4.61 billion to $4.78 billion

Subsidies to capital costs

According to the DOE, government ownership of the MOX fuel fabrication facility saves the government approximately $600 million. This is due to the lower cost of capital relative to private financing, no interest during construction, and no need for a rate of return for private companies. Thus, if the cost of MOX and LEU fuel were compared on a consistent financing basis, the cost differential would be $620 million, not the $20 million as shown in the Net Life Cycle cost for MOX fabrication shown for LWR's in Table 6.2. For comparison, a new column of data for private financing has been added to Table 6.2 to put all costing on an equal basis.

The federal funds expended to avoid paying the higher interest rates that accompany the use of private capital are subsidies. They are defined this way because the government funds could have been lent out or used to reduce the debt, and because this benefit is not available to competing, non-nuclear power producers. The value of this subsidy is the difference between the cost of private capital and cost of using government funds.

Existing Reactors: $0 million to $750 million
Evolutionary and Partially Completed Reactors: $0 million

Total life-cycle subsidies required by a MOX user - the sum of all subsidies in all categories

Existing Reactors: $3.2 billion to $4.85 billion
Evolutionary and Partially Completed Reactors: $7.7 billion to $11.7 billion

Offsets to subsidy costs - fuel displacement credits

If a nuclear power producer is competitive with other forms of power production, and if that producer elects to use MOX that has been provided free by the DOE, the federal government should be reimbursed for the amount of LEU fuel that the MOX has replaced. This reimbursement is not a payment for the MOX fuel, it is simply a reflection of costs of operation offset by the use of MOX. If the price of uranium rises, or if the cost of other factors involved in LEU fuel fabrication increase, and if this is reflected in an increase in the cost of LEU fuel, then the size of the fuel displacement reimbursements should also rise. At current LEU prices, the range of these credits would be between:

Existing Reactors: $320 million to $2.01 billion
Evolutionary and Partially Completed Reactors: none

On the other hand, if the nuclear power operator is not competitive with alternative forms power production, it may reduce fuel displacement reimbursements below the actual amounts of LEU fuel displaced. The amount that a power producer falls short of reimbursing the federal government for displaced fuel would also be a subsidy to the power producer. As competitiveness decreases, either reimbursements will decrease - and subsidies will increase - to the point where no fuel displacement reimbursements are being returned to the federal government by the power producer, or the power producer will be forced to incur increasing losses. In such a scenario, the producer would either go out of business, or the federal government would be forced to provide additional subsidies. While this eventuality may seem problematic, it is fair to note that the federal government is unlikely to allow a power producer to fail if that producer has become a critical part of a plutonium disposition program involving MOX burning.

Total subsidies required by MOX users with credit for fuel displaced

It has often been claimed that even if MOX was free, nuclear power producers would prefer to buy LEU fuel. The following figures show why this statement is true. Irrespective of the fuel displacement credit generated by MOX use, the life-cycle costs of MOX still greatly exceed the costs of LEU. The ranges of subsidy required by existing reactor MOX users - after full reimbursement for fuel displacement credits - would be:

Existing Reactors - Low Range (Govt. Financing) $1.92 billion to $3.11 billion
Existing Reactors - High Range (Private Financing) $2.09 billion to $3.71 billion

For evolutionary and partially completed reactors, the revenues shown in Table 6.2 accrue to the owners of the plant, not to the government. Thus, their impact on the plutonium disposition mission costs is unknown (Technical Summary Report For Surplus Weapons-Usable Plutonium Disposition, 1996, p. 4-7). In spite of this, DOE showed these revenues as reductions in government costs. To correct this, Table 6.2 also includes the costs for the more likely option where the federal government does not receive these revenues. There is no indication that any of these revenues will be returned to the federal government - unless the government becomes the owner of the plant. For this reason, the subsidy amounts required from the federal government remain constant at:

Evolutionary and Partially Completed Reactors: $7.7 billion to $11.7 billion

Additional MOX subsidy issues

As the price of LEU rises and it becomes less competitive with other sources of energy, the subsidy provided to the MOX burner must cover not only the difference between the price of LEU and the price of MOX, but also the difference between the price of LEU and the price of the most economically competitive energy source. In every case the subsidy required to get an energy producer to use MOX must be the

84

difference between the cost of MOX and the cost of a fuel source that is most economically competitive. However, if such a subsidy was actually delivered to a nuclear power producer, it could potentially make MOX burning economically viable in a market where no form of nuclear energy is competitive. These subsidies have the potential to "save" nuclear power producers who cannot compete in a deregulated environment.

In its report on surplus weapons-usable plutonium disposition DOE included no calculations of 'incentive fees' and noted that this created "a significant cost uncertainty" in DOE's cost estimates. (Technical Summary Report For Surplus Weapons-Usable Plutonium Disposition, 1996, p. 4-3) This 'cost uncertainty' springs from at least four different problems that arise when a government agency elects to intervene in the operation of a market by subsidizing one sector:

1.   The difficulty of accurately calculating an appropriate level of subsidy to exactly compensate for MOX use and the danger that such a subsidy, if miscalculated, would create a revenue base and competitive pricing advantage for nuclear power producers that would drive other, naturally competitive power producers out of the market.
2.   The known inefficiencies of involving a government agency in the operations of a private institution.
3.   The inability of the federal government, which budgets on an annual or bi-annual basis, to react to day-today changes in market pricing resulting from price changes in other fuels, shifts in the customer base, or changes in related federal regulations in a timely or knowledgeable manner.
4.   The implication, given the amounts of plutonium requiring disposition, that subsidies would be a twenty to thirty year guaranty, while federal budgeting methods cannot guarantee funding for more the one or two years. This would increase the financial risk to power producers - and any increased risk would have to be reflected in higher subsidies.

Other estimates of MOX facility and MOX use costs

Historically, all credible estimates of the cost of making and using MOX have exceeded the cost of using LEU in LWRs. Further, while most cost estimates have anticipated a surplus of civilian plutonium in the foreseeable future, few have found ways to incorporate all the costs associated with the proliferation concerns MOX raises. The MOX fuel for a single 1 gigawatt LWR contains enough plutonium for 70 bombs (Chow and Solomon, 1993, pp. xvi, xvii). Table 6.3 lists various MOX facility and MOX burning cost estimates. The validity of any of these estimates is questionable, but the relatively small range of cost estimates for a given type of facility is instructive.

Table 6.3
Recent estimates of MOX facility and MOX burner costs
millions of 1996 dollars

Source	Type	Date	Cost	Time to Construct
Fabrication Facilities:				
Schulze	MOX Facility	1992	300	5 years**
Westinghouse SRS-New Site 60 MT/yr.	MOX Facility	1992	682	Not Given***
NAS	MOX Facility	1994	400-1200	10 years****
DOE Existing Site	MOX Facility	1996	400	10 years*
DOE New Site	MOX Facility	1996	450	10 years*
Diakov	MOX Facility	1996	250	Not Given^
MOX Burners:				
vonHipple et al.	ALMR	1993	$Billions for Development	10 years^^
vonHipple et al.	MHTGR	1993	$Billions for Development	10 years^^

Sources:
*Technical Summary Report For Surplus Weapons-Usable Plutonium Disposition (1996), pp. 4-7.
**Schulze (1992), pp. 65-74.
*** Buckner et al. (1992), p. 10.
**** Management and Disposition of Excess Weapons Plutonium (1994), pp. 159-160.
^Diakov (1996).
^^ von Hippel et al. (1993), p. 48.

Capacity and locations of MOX facilities

The world's LWRs produce about 70 tons of plutonium per year. Existing plants for making MOX have a capacity of about 88 tons/year and construction of additional capacity of 260 tons/year is under way in France, Japan and Germany. An additional 120 tons/year capacity is being built in Russia. When all construction is completed, and assuming the plants function as designed, there will be a potential world capacity of 468 tons of MOX. About 2000 tons of uranium are required to mix with 70 tons of plutonium to make MOX - a ratio of roughly 25 parts uranium to 1 part plutonium. Thus, 200 tons of military plutonium would require about 5000 tons of uranium mix to make MOX (Garwin, 1992, pp. 17-20).

Based on the above ratio, the world's MOX capacity, when constructed, could handle 18.7 tons (468/25) of plutonium - both weapon and reactor grade - per year. If all planned MOX capacity was fully operational and entirely devoted to getting rid of weapon-grade plutonium, it would take 11 years to convert the current 200 tons of weapons grade plutonium to MOX. Thus, the MOX capability - either existing, planned or under construction - is totally inadequate to handle either the spent fuel or the surplus of military plutonium, and storage will be necessary for extended periods of time.

The estimated cost for new MOX fuel is about $1450-$1800/kg in 1996 dollars. At these MOX prices it would cost about $5.6 billion to dispose of 200 tons of plutonium. This cost could range from $2.2-$11 billion in 1996 dollars using a reasonable range of future LEU and MOX prices. Of course, burning plutonium in MOX fuel does not completely eliminate it. About 300 kg of new plutonium is produced for each metric ton of plutonium loaded in the reactor (Fetter, 1992, pp. 144-148).

Fabrication of MOX from surplus weapon plutonium - special problems

Planning and cost estimates concerning MOX fabrication are generally based on the experience of civilian MOX fabricators who have used reprocessed plutonium from spent reactor fuel. However, if one chooses to dispose of excess weapon grade plutonium by making it into MOX, a substantial problem arises concerning the gallium used as an alloying element in the plutonium pits of nuclear weapons.

MOX is a ceramic substance prepared by sintering oxides of plutonium and uranium. Concentrations of gallium affect the sintering behavior of ceramics and gallium chemically attacks zirconium, a metal used in cladding nuclear fuel. For both these reasons, gallium must be removed from surplus weapon plutonium before it can be used in MOX. This would require either aqueous reprocessing which creates large quantities of liquid wastes or removal after the plutonium is in an oxide form as part of the proposed 'Aries' process for pit disassembly. For obvious reasons, an aqueous process is not desirable and the resulting waste stream would significantly increase the cost of the MOX produced. However, the 'Aries' process has only been demonstrated in a laboratory setting and the process for removing gallium is not yet fully developed (Toevs and Beard, 1996). As a result, the DOE has estimated that gallium removal from surplus weapon plutonium could increase the costs of MOX fabrication covered previously in this chapter by about $200 million (Technical Summary Report For Surplus Weapons-Usable Plutonium Disposition, 1996, p. 6-3).

Civilian MOX fabrication

In 1996, there were three MOX fuel assembly fabrication plants in operation in the world: Companie Generale des Matieres Nucleaires (COGEMA) Complexe de Fabrication Des Combustibles Plant located at Cadarache(France), Etablissement MELOX Plant at Marcoule (France), and Belgonucleaire Usine de Fabrication D'Elements PU Plant at Dessel (Belgium). Specific capacities and completion schedules for MOX plants are as follows:

France

Plutonium was first separated in France in 1949. Today, the plutonium content in the 1300 tons of spent fuel created annually by French operations is about 10 tons, but only 850 tons of spent fuel are reprocessed each year - yielding about 8.5 tons of spent fuel. By the end of 1994, 460 MOX fuel assemblies had been supplied by COGEMA to seven French PWRs and to customers abroad. In 1995, the new 120 THM/yr. MELOX plant at Marcoule started industrial fabrication of MOX. This plant could easily be upgraded to 160 THM/yr. and construction of another 120 THM/yr. plant is planned at LaHague (Decressin et al., 1996, p. 3).

France's Marcoule (MELOX) plant, with a capacity of 115 tons, can handle about 3.5 tons of plutonium per year (Schulze, 1992, pp. 65-74). The MELOX plant had produced 76 fuel elements by the end of 1995. It is expected that 160 assemblies will be fabricated in 1996. The MOX fabrication facilities at Cadarache have been in operation since 1963, mainly for producing fuel for fast reactors, but the plant has been engaged in fabricating MOX fuel for light water reactors(LWR) since 1990. The capacity of Cadarache is about 15 tons/year, and more than 30 tons of plutonium have been processed since the plant opened (Nigon and Fournier, 1996). A planned facility at Meloux will have a capacity of 120 tons, and could handle about 4 tons of plutonium per year (Vendryes, 1992, pp. 61-64).

Germany

Siemens had an older, small demonstration facility in operation, but it was closed down in 1991 after a contamination accident and an inability to meet the safety standards prescribed in the new German Atomic Energy Law (Kueppers and Sailer, 1994, Ch. 4).

A new 120 ton German MOX plant with a capacity of 5 tons of plutonium per year has been completed by Siemans (Schulze, 1992, pp. 65-74). However, the plant received only a partial license in 1987 when it started construction. In 1993, the Higher Administrative Court declared three of the partial licenses to be unlawful (Kueppers and Sailer, 1994, Ch. 4). By April, 1994, it was apparent that it was impossible to gain approval of the state government of Hesse for operation of the new plant for both safety and economic reasons. As a result, utilities decided to neither finance completion of the new plant nor to keep paying maintenance costs.

Siemens abandoned the plant in 1995 and declared themselves "forced to abandon MOX fuel production in Germany." (Siemens Company, 1995)

Belgium

The BELGONUCLEAIRE plant (P0) has a capacity of 35 tons per year, or about 1.2 tons of plutonium (Schulze, 1992, pp. 65-74). This plant has been in operation since 1973, mainly fabricating MOX fuel for fast breeder reactors. MOX for LWRs in Belgium, France, Germany and Switzerland has been commercially produced since 1984. By the end of 1995, more than 270 tons of MOX fuel assemblies had been manufactured at this plant (van Vliet et al., 1996). BELGONUCLEAIRE has a project to extend the Po plant by adding two new production lines with the capacity of 40 tons per year. Design of the plant has been completed, but due to licensing problems, the plant has never been constructed (Ayukawa, 10/3/1996).

Canada

Canada's deuterium-uranium reactor (CANDU) is able to burn MOX in full core loads and does not have to be shut down to refuel. This allows more plutonium to be consumed, but a standard CANDU can still only use one metric ton of plutonium a year (Charnctski and Rauf, 1994).

Great Britain

The Sellafield MOX Plant (SMP), a large-scale commercial MOX fuel fabrication plant, is currently under construction in the UK. This plant should start operation in 1997 with an annual capacity of 120 metric tons. However, since Britain gave up its fast breeder reactor projects in 1994, and since there is no ongoing MOX program in Britain, this plant has been constructed solely to create export services to BNFL's foreign reprocessing customers, notably Japan and Germany (Ayukawa, 10/3/1996).

Japan

A research-scale fabrication plant, the Plutonium Fuel Production Facility, is operated by the government-owned Power Reactor and Nuclear Fuel Development Corporation located in Tokaimura. It has produced MOX fuel for fast breeder reactors and demonstration assemblies for light water reactors. In May, 1994, Nuclear Control Institute revealed that 70 kg of plutonium was lost in the plant's systems. The Japanese government admitted that this problem had been identified by IAEA, but that it had been unable to recover the plutonium (Nuke Info Tokyo, 1995).

Russia

Russia began experiments with MOX fuel fabrication in the mid 1950's. Three pilot plants produce MOX at the present time, two at MAYAK and one at Dimitrovgrad. These plants are not modern and their use for plutonium disposition is doubtful. Construction of an industrial-scale MOX plant was started at MAYAK and then suspended due to lack of funds. This plant was supposed to generate 900 fuel assemblies or 60 MT of MOX fuel annually. The cost of a demonstration MOX plant in Russia is estimated to be $60 million in 1996 dollars, with the cost of a full production plant estimated at about $250 million (Diakov, 1996).

United States

The US currently has no MOX plants but some parts of the DOE would like to construct MOX facilities for use in disposing of excess weapon-grade plutonium. At a meeting held for citizens to discuss this issue with the DOE, DOE suggested that the TA55 facility at Los Alamos National Laboratory be used for plutonium processing (pit disassembly). For MOX fuel fabrication, the DOE suggested the Fuel and Materials Examination Facility at Hanford, an empty building at the Idaho National Engineering Laboratory, a new building at the Pantex plant, and either the New Special Recovery Facility or the F Canyon Facility at the Savannah River Site (Eldredge, 1996). Recent reports have claimed that between 40 and 60 of the 110 commercial US LWRs are suited to the use of MOX (Rougeau, 1996, Ayukawa, 9/23/1996).

MOX utilization

In 1996, France, Germany, Switzerland and Belgium were the only countries actually using MOX fuel in LWRs. Because LWRs were originally designed to burn uranium fuel and MOX fuel has different nuclear characteristics, LWRs must be re-licensed to burn MOX. Since the experiences gained through actual MOX use are being analyzed and evaluated, the 18 reactors currently burning MOX are ongoing research and development projects for MOX utilization in LWRs (Ayukawa, 10/3/1996).

France

Of the 54 LWRs in operation in France today only sixteen reactors at Saint-Laurent-des-Eaux, Gravelines, Dampierre, Blayais, and Tricastin are licensed to use MOX fuel assemblies (Kueppers and Sailer, 1994, Ch. 5). Of these, MOX fuel has been loaded in seven reactors to date: Saint-Lauren B2(1988), Gravelines 3&4(1989), Dampierre 1(1990), Dampierre 2(1993), and Blayais 2(1994) (Provost, 1996). In addition, the first shipment of spent MOX fuel was delivered to La Hague in early 1996.

Currently, one third of a reactor core is loaded with MOX and the MOX fuel is irradiated for three years before it is replaced. However, insufficient plutonium is used by this process to consume the amounts of plutonium available. For this reason, Electricite de France has proposed to license additional reactors to burn MOX and to load reactors with 50% MOX fuel (Provost, 1996).

Germany

Germany has the most ambitious MOX program. After Germany canceled the Kalkar fast breeder reactor in 1991, the 25 tons of plutonium now in the German stockpile was scheduled to be used in the seven reactors now operating with partial MOX fuel. Five more reactors have been licensed to start loading MOX fuel, and three more are applying for a license (Kueppers and Sailer, 1994, Ch. 5). However, public acceptance of nuclear power has deteriorated, and with the revision of the Atomic Energy Law, it is virtually impossible to further expand the MOX program. As a result, German utilities canceled some of the contracts they had with BNFL in 1994.

Switzerland

Switzerland has five reactors in operation and has been using MOX fuel assemblies in two of them since 1978. Switzerland plans to receive 2.2 tons of plutonium from its reprocessing contracts by the year 2003. Previously, MOX fuel assemblies have been supplied by Belgonucleaire, Siemens and BNFL. Swiss nuclear policy permits all five of its reactors to burn MOX fuel, and all reactors are licensed to load MOX fuel assemblies up to 40% of the reactor core (Stratton and Bay, 1996).

Belgium

Belgium is currently the leading country in MOX fuel fabrication. However, Belgium decided in 1994 to make no further reprocessing contracts, freeze post-2000 contracts, and use 4.6 metric tons of separated plutonium from older contracts as MOX fuel in Belgium reactors (van Vliet et al., 1996). Licenses to load these MOX fuel assemblies were granted in November, 1994. In 1995 two Belgium reactors were burning MOX fuel (Synatom, 1995).

Japan

According to the long-term Program revised in 1994, Japan expects 70-80 tons of plutonium to be separated by the year 2010. Of this amount, 25 tons are to be used in fast breeder reactors and advanced thermal reactors (Long-Term Program for the Development and Utilization of Nuclear Energy 1994, 1994). The remaining 50 tons are to be burned as MOX in LWRs. However, in August, 1995 the new advanced thermal reactor planned for Ohma was canceled for economic reasons and the Monju fast breeder reactor was shut down because of a sodium leak. As a result, almost all plutonium was being burned in LWRs (Ayukawa, 10/3/1996).

Plutonium consumption rates in MOX

The first step in creating MOX from weapon-grade plutonium is to convert the plutonium into plutonium dioxide. The process to do this does not yet exist (Rougeau, 1996). However, such a process is under development in the US and it has been given the tentative name of ARIES - Automated Recovery and Integrated Extraction process.

Plutonium's nuclear characteristics limit the amount of plutonium that can be used in LWRs to about one third of the core. As a result, a 1,000 megawatt reactor could use only about 300 kg of plutonium a year. Russia, with seven such reactors in operation, could only irradiate about 2.5 tons of plutonium a year. At this rate, it would take 40 years to irradiate Russia's 100 tons of weapon-grade plutonium (von Hippel et al., 1993, p. 48).

A regular uranium dioxide (UO_2) reactor fuel assembly contains no plutonium when loaded in a commercial reactor. After irradiation, it is about 1% plutonium. Thus, a 500 kg fuel rod contains about 5.5 kg of plutonium by the time it has been converted to spent fuel. Fresh civilian MOX fuel contains roughly 4% plutonium when it has been fabricated for LWRs and roughly 20% plutonium when it has been made for fast reactors. For light water reactors, of the 20 kg of plutonium loaded into fuel assemblies, about 15 will remain after irradiation. Depending on the composition of the MOX, a LWR loaded with one third of its core in MOX will consume an amount of plutonium in the MOX assemblies that roughly equals the amount of plutonium created in the remainder of the uranium-based reactor fuel (Rougeau, 1996). The results of this approach to plutonium burning are approximately:

6 MT Pu => 150 MT MOX (4%Pu) => 8 LWRs (Full Core Load)
24 LWRs (1/3 Core Load) => 150 tons spent fuel (~ 2.5% Pu)

The plutonium remaining in the spent fuel would have an increased fraction of plutonium isotopes other than Pu-239, making the material less attractive for weapons. However, plutonium from spent fuel could still be chemically separated and used to make simple bombs (von Hippel et al., 1993, p. 47). Further, the process has created 150 tons of spent fuel that must be handled, treated and disposed of at costs that all decrease the economic benefit of reactor use.

More plutonium would be consumed and security risks could be lowered if reactors could take a full core load of MOX. A LWR designed by ABB Combustion Engineering has been designed to do this, but such a reactor remains in the theoretical realm. Other candidates for MOX burning include the Liquid Metal Fast Neutron reactor (ALMR), and the modular high temperature gas cooled reactor (MHTGR). The fast neutron reactor can irradiate more plutonium than a LWR of similar power, but the spent fuel is still near weapon-grade plutonium. The results of using the ALMR would be:

6 MT Pu => 30 MT MOX (20% Pu) => 3 Fast Neutron Reactors => 30 tons spent fuel (20% Pu)

The Plutonium's nuclear characteristics limit the amount of plutonium that can be used in LWRs to about one third of the core. reactor could irradiate plutonium to a point where most of it was destroyed and the remainder residing in spent fuel was more undesirable than that from a LWR (von Hippel et al., 1993, p. 48). However, the central issue is still how much spent fuel remains to be dealt with and disposed of after the plutonium disposition method has been employed. As a point of comparison with these options, direct disposal through vitrification would create the following amounts of waste:

6 MT Pu + 10's of Megacuries of Cesium 137 and other fission products => Vitrification Facility => 100-500 tons Steel-encased radioactive glass logs (von Hippel et al., 1993, pp. 48-49).

According to the recent DOE report on surplus weapons-usable plutonium disposition, using vitrification to dispose of all surplus weapons plutonium would only cost about $1.8 billion (Technical Summary Report For Surplus Weapons-Usable Plutonium Disposition, 1996, p. 4-10).

References

'Arizona Public Service Company Letter to Tom Clements, Greenpeace International, September 3, 1996' (1996), reported in Ayukawa, Yurika, *Fissile Material Disposition & Civil Use Of Plutonium*, Issue No. 1, yayukawa@igc.apc.org.

Ashton, Jack (1996), Editor/Contact, *Uranium Institute, 1996 NucNet*, nucnet@atagbe.ch, August 26.

Ayukawa, Yurika (10/3/1996), *Fissile Material Disposition & Civil Use Of Plutonium*, Issue No. 2, yayukawa@igc.apc.org.

Ayukawa, Yurika (9/23/1996), *Fissile Material Disposition & Civil Use Of Plutonium*, Issue No. 1, yayukawa@igc.apc.org.

Berkhout, Franz (1993), *Fuel Reprocessing At THORP: Profitability and Public Liabilities*, Center for Energy and Environmental Studies, Princeton University, Princeton, NJ, p. 6.

Buckner, M.R., Radder, J.A., Angelos, J.G., Inhaber, H. (1992), *Excess Plutonium Disposition Using ALWR Technology*, WSRC-RP-92-127B, Westinghouse Savannah River, Aiken, SC, p. 10.

Charnetski, Joanne and Rauf, Tariq (1994), 'Let Canada Cremate Nuclear Swords', *Defense News*, October 3-9.

Chow, Brian G. and Solomon, Kenneth A. (1993), *Limiting the Spread of Weapon-Usable Fissile Materials*, National Defense Research Institute, RAND, Santa Monica, CA, pp. xvi, xvii, 14.

Cumo, Maurizio (1992), 'General Economic Evaluation of the Conversion of Nuclear Warheads into Electrical Power', *Working Papers of the International Symposium on Conversion of Nuclear Warheads for Peaceful Purposes*, Rome, Italy, pp. 95-97.

Decressin, A., Gambier, D.J., Lehmann, J., Nietzold, D.E. (1996), *Experience and Activities in the Field of Plutonium Recycling in civilian Nuclear Power Plants in the European Union*, Paper Presented at the International Conference on Military Conversion and Science: Utilization/Disposal of the Excess Weapon Plutonium: Scientific, Technological and Socio-Economic Aspects, Como, Italy, p. 3.

Diakov, Anatoli S. (1996), *Utilization of Already Separated Plutonium in Russia: Consideration of Short- and Long-Term Options*, Paper Presented at the International Conference on Military Conversion and Science: Utilization/Disposal of the Excess Weapon Plutonium: Scientific, Technological and Socio-Economic Aspects, Como, Italy.

Eldredge, Maureen (1996), 'Notes From The September 10, 1996, DOE Briefing', Military Production Network, in Ayukawa, Yurika, *Fissile Material Disposition & Civil Use Of Plutonium*, Issue No. 1, yayukawa@igc.apc.org.

Fetter, Steve(1992), 'Control and Disposition of Nuclear Weapons Materials', *Working Papers of the International Symposium on Conversion of Nuclear Warheads for Peaceful Purposes*, Rome, Italy, pp. 144-148.

Fuoto, John, Ogden Environmental And Energy Services, personal communication to William J. Weida, jsfuoto@oees.com, (January 27, 1997).

Garwin, Richard L. (1992), 'Steps Toward the Elimination of Almost All Nuclear Warheads', *Working Papers of the International Symposium on Conversion of Nuclear Warheads for Peaceful Purposes*, Rome, Italy, pp. 17-20.

Kerber, Ross (1996), 'Nuclear-Power Plant Shutdown 'Likely' For Facility Run by Northeast Utilities', *The Wall Street Journal*, October 10, p. A6.

Kueppers, Christian and Sailer, Michael (1994), *The MOX Industry or The Civilian Use of Plutonium*, International Physicians for the Prevention of Nuclear War, Chapters 4, 5.

'Long-Term Program for the Development and Utilization of Nuclear Energy 1994' (1994), *Nuke Info*, No. 41, Tokyo, Japan.

Management and Disposition of Excess Weapons Plutonium (1994), Committee on International Security and Arms Control, National Academy of Sciences, National Academy Press, Washington, DC, pp. 159-160.

National Conference of State Legislatures (1996), *High-Level Radioactive Waste Newsletter*, July.

Nigon, J.L. and Fournier, W. (1996), 'MOX Fabrication and MOX Irradiation Experience Feedback from the French Programme', COGEMA, *International Seminar on MOX Fuel: Electricity Generation from Pu Recycling*, United Kingdom.

Nuclear Fuel (1996), September 9, p. 2.

Nucleonics Week (1996), April 4.

Nuke Info Tokyo (1995), Citizens' Nuclear Information Center, Japan, No. 50.

Numark, Neil J. (1996), *Get SMART: The Case for a Strategic Materials Reduction Treaty, and Its Implications*, Paper Presented at the International Conference on Military Conversion and Science: Utilization/Disposal of the Excess Weapon Plutonium: Scientific, Technological and Socio-Economic Aspects, Como, Italy, p. 6.

Provost, J.L. (1996), *Plutonium Recycling and Use of MOX Fuel In PWR - French Viewpoint*, Presented at the International Seminar on MOX Fuel: Electricity Generation from Pu Recycling, United Kingdom.

Rougeau, Jean-Pierre (1996), *A Clever Use of Ex-Weapons Material*, Paper Presented at the International Conference on Military Conversion and Science: Utilization/Disposal of the Excess Weapon Plutonium: Scientific, Technological and Socio-Economic Aspects, Como, Italy.

Schneider, Mycle (1996), *Plutonium Fuels*, World Information Service on Energy, Krasnoyarsk, Russia, p. 4.

Schulze, Joachim (1992), 'Burning of Plutonium in Light Water Reactors (MOX Fuel Elements) Compared To Other Treatment', *Working Papers of the International Symposium on Conversion of Nuclear Warheads for Peaceful Purposes*, Rome, Italy, pp. 65-74.

Siemens Corporation (1995), *Press Release*, July 7.

Silver, R. (1996), 'Hydro Puts Off Bruce Retubing As Hope for Pu Mission Fades', *Nucleonics Week*, August 15, p. 8.

Stratton, R. and Bay, H. (1996), *Experience in the Use of MOX Fuels in the Beznau Plants of NOK*, Presented at the International Seminar on MOX Fuel: Electricity Generation from Pu Recycling, United Kingdom.

Synatom (1995), '1995 Report'.

Technical Summary Report For Surplus Weapons-Usable Plutonium Disposition (1996), DOE/MD-0003, Office of Fissile Materials Disposition, US Department of Energy, Washington, DC, pp. 4-3 to 4-10, 6-3.

Toevs, J. and Beard, C. (1996), *Gallium in Weapons-Grade Plutonium and MOX Fuel Fabrication*, LA-UR-96-4764, Los Alamos National Laboratory, NM.

van Vliet, J., Haas, D., Vanderborck, V., Lippens, M., Vandeberg, C. (1996), *MIMAS MOX Fuel Fabrication & Irradiation Performance*, Belgonucleaire, International Seminar on MOX FUEl: Electricity Generation from Pu Recycling, United Kingdom.

Vendryes, Georges (1992), 'Plutonium Burning In Fast Reactor and As MOX Fuel', *Working Papers of the International Symposium on Conversion of Nuclear Warheads for Peaceful Purposes*, Rome, Italy, pp. 61-64.

von Hippel, F., Miller, M., Feiveson, H., Diakov, A., Berkhout, F. (1993), 'Eliminating Nuclear Warheads', Scientific American, August, pp. 47-49.

7 Burning plutonium in reactors: implications for disposition

Introduction

The use of plutonium in reactors has been proposed for two reasons - first, to use the energy in plutonium to generate power (80 tons of plutonium contain the energy equivalent of about 160,000,000 tons of oil) (Vendreyes, 1994) and second, to make plutonium unusable as weapon-grade material. Because light water reactors (LWRs) can only be partially loaded with plutonium, destruction of plutonium in these reactors is a slow process. Almost as much plutonium is created from the uranium fuel in the reactor core as is destroyed in the plutonium-based fuel.

For example, the French have stated that when plutonium is burned in their light water reactors roughly two-thirds of the amount consumed is reproduced from the uranium in the reactor core. As a result, the actual rate at which plutonium disappears is on the order of only 300 kg per year for a 1000 MWe reactor. In addition, the quality of plutonium as LWR fuel decreases continuously as irradiation continues (Vendreyes, 1992, pp. 61-64). Table 7.1 illustrates the small rate of plutonium consumption in a variety of Russian LWRs and in two advanced Russian reactors (the BN-600 and BN-800) where full core loads of plutonium are possible. Note that for light water reactors attempts to burn plutonium results in more plutonium after the burning is completed.

It is not desirable to store MOX fuel for long periods because reactor plutonium progressively turns into radioactive Americium 241 (Decressin et al., 1996, p. 7). Americium 241, with a half life of 14.4 years, is a strong alpha and gamma emitter and must be limited in concentration. This is not a problem with weapon-grade plutonium because of its purity, and long term storage of weapon-grade material is possible (weapon grade plutonium has about 94% Pu-239 while reactor grade plutonium has about 60%.) Tests have also shown that mixing weapon-grade Pu-239 with Pu-238 is not practical. Instead, to make MOX, all plutonium must first be converted to plutonium dioxide (Schulze, 1992, pp. 65-74). Finally, as the US demonstrated in the 1960's, reactor-grade plutonium can still be used to construct bombs.

97

Table 7.1
Plutonium consumption rates for Russian reactors

	Plutonium:		
	Loading	Yield	Balance
Reactor	kg/yr.	kg/yr.	kg/yr.
1. VVER-1000	0	223	+223
2. VVER-1000	254	308	+54
3. VVER-1000	364	395	+31
4. BN-600	1141	1053	-88
5. BN-800	1637	1508	-129

Notes:
1: reactor loaded with LEU only.
2: 1/3 reactor loaded with MOX with 3.5% Pu.
3: 1/3 reactor loaded with MOX with 5% Pu.
4, 5: reactor fully loaded with MOX.
Source: 'Engineering analysis of production of uranium/plutonium fuel from weapon plutonium and its possible utilization in nuclear energentics' (1995), MINATOM of Russia, Siemens, GRS, in Bolshov, L.A. (1996), 'Environmental Safety and Health Risks of the Different Plutonium Disposition Options', Paper Presented at the International Conference on Military Conversion and Science: Utilization/Disposal of the Excess Weapon Plutonium: Scientific, Technological and Socio-Economic Aspects, Como, Italy, March 18-20, Slide 4.

Based on Table 7.1 and on the French experience, there is little possibility of burning large amounts of weapon-grade plutonium in fast breeder reactors. Since the Russian BN-600 and BN-800 reactors combined could only burn about 220 kg of plutonium a year, they would take over four years to destroy a single ton of plutonium. The performance of other fast reactors is not much better. A 1992 study of plutonium consumption in Japanese reactors by Uematsu estimated that the MONJU reactor - a 280 MWe fast neutron reactor - could burn an average of about 210 kg/year of plutonium while about 130 kg/year of plutonium would be produced in the core. Thus, the net burn up for this type of reactor was only about 38% of the plutonium per year (Uematsu, 1992, pp. 109-115).

Since existing fast breeder reactors can only burn a limited amount of plutonium, any effective program to dispose of plutonium through burning would require a massive reactor construction program - a very expensive proposition. Uematsu estimated the total demand for plutonium for all reactor use between 1995 and 2010 is only about 186.3 tons. Over the same 15 year period, another 345 tons of new civilian plutonium will be produced by normal reactor operations (Uematsu, 1992, pp. 109-115). As previously noted, civilian plutonium decays to Americium 241 which makes it more difficult to handle. And since plutonium isotope quality also decays in the presence of the radiation, plutonium cannot be recycled more than once or twice, leaving a substantial amount of material to be dealt with after the recycling is done (Vendreyes, 1994).

Developmental and alternative technologies for burning plutonium - breeder reactors

One answer to the small burn up rates experience by most conventional reactors is to use a different reactor design that can accommodate a full core load of plutonium and thus increase the burn rate. Three US reactors of the System 80 type at the Palo Verde Nuclear Generating Station are pressurized light water reactors (PWRs) that could technically handle a full core load of MOX. Using these reactors, it would take 30 reactor years - or 10 years for all three reactors - to convert 50 tons of plutonium into spent fuel (Makhijani and Makhijani, 1995, pp. 26-27). However, most proponents of plutonium use in reactors propose to use some variation of a breeder reactor to maximize plutonium consumption.

Proponents of plutonium use in breeder reactors acknowledge that these reactors are currently very uneconomical. However, they also claim that breeder reactors will be necessary in the future when other fuels become scarce. The United States spent about $3.3 billion in 1996 dollars on breeder reactor technology in the 1970's and early 1980's when uranium was expensive and when the number of nuclear power plants in the United States and abroad was expected to continue to increase. Economics and politics combined to undermine the nuclear industry and partly as a result, the price of uranium, which was $40 a pound in 1982, had dropped to $16 a pound in 1996 (Wald and Gordon, 1994).

von Hippel has calculated that plutonium breeder reactors will not be economically competitive for nearly a century. Non-competitiveness was the main reason Congress terminated the Clinch River plutonium-breeder reactor in 1983 (Bradley, 1993). Advocates had tried to sell the Clinch River Breeder Reactor (CRBR) as a source of power, but power companies refused to finance the project because its costs had skyrocketed - rising from $257 million to $8 billion. As a result, Congress canceled the CRBR and many observers feel that similar economic impediments will delay the use of fast reactors until at least 2030/2050 (Lombardi, 1994).

Advanced liquid metal reactors

Following the demise of the Clinch River project, breeder reactor proponents developed the Advanced Liquid Metal Reactor (ALMR). The National Academy of Sciences has reported that the ALMR's ability to consume plutonium is "not sufficient to greatly alter the security risks posed by the material remaining in the spent fuel." It further added that ALMRs do "not offer sufficient advantages to overcome their liabilities of cost, timing and uncertainties." (Lindsay, 1994)

Throughout the many lives of the ALMR, three things have remained relatively consistent:

1.   There has been no real commercial interest in the reactor.
2.   The environmental hazards of the process have been generally agreed upon.
3.   A significant proliferation concern has surrounded the ALMR's operations (Lindsay, 1994).

By the late 1980's, the ALMR had, in its latest reincarnation, become the Integral Fast Reactor (IFR). The IFR consists of a liquid metal-cooled fast reactor coupled with a compact pyroprocessor for reprocessing. The IFR can also be used as a breeder reactor. Arguments by proponents for the IFR have centered around the following points:

1.   The IFR can burn spent fuel or plutonium, generating electricity while reducing the volume and radioactivity of nuclear waste (Lippman, 1993).
2.   IFR has routinely produced three times as much energy as present day reactors, with experiments demonstrating factors of six times more (Integral Fast Reactor, 1993).
3.   The IFR can be operated as a plutonium burner, as a "break-even" reactor that is not a net producer or burner of plutonium, or as a breeder that produces more plutonium than it burns (Protection and Management of Plutonium, 1995, p. 14).
4.   Each time fuel passes through the IFR about 20% of the initial amount of fuel is used to produce power. Conventional reactors use only about 0.5% of the energy available in the uranium (Vargas, 1993).
5.   Waste generation is minimized through innovative technology, aggressive reprocessing of chemicals and nuclear material, and elimination of most hands-on maintenance of radioactive components (Integral Fast Reactor, 1993).
6.   Other countries will have more confidence in the irreversibility of the warhead reduction if reactor irradiation is used to convert plutonium to the spent fuel standard (Protection and Management of Plutonium, 1995, p. 10).

Opponents to the IFR have countered these claims by noting that:

1.   The liquid sodium in the IFR is highly flammable. A 1995 sodium leak at MONJU, Japan's prototype FBR, burned, reacting with moisture in the air, and as much as five cubic meters of sodium poured onto a floor of the reactor and solidified (Takagi, 1995).

2. It makes no sense to reprocess unless you intend to have a breeder program produce plutonium as fuel. You can make new uranium fuel for a fraction of the cost of plutonium fuel (Lippman, 1993).
3. Reusing the same fuel creates reprocessing waste each time reuse is attempted (Integral Fast Reactor, 1993).
4. Running the IFR as a plutonium reducer does not require fuel reprocessing, but it produces large quantities of intensely radioactive waste. When used in a reprocessing 'waste reduction mode', the IFR produces more waste than it consumes (Integral Fast Reactor, 1993).
5. The IFR does not burn up waste. It transforms spent fuel into highly radioactive waste that is no less dangerous, but doesn't have to be classified as high level waste simply because it is no longer spent fuel (Integral Fast Reactor, 1993).

To date, these concerns have far outweighed the proposed benefits of the IFR. As support for the IFR wound down, Senator J. Bennett Johnston presented what appeared to be the only remaining economic argument for the reactor:

> We have already spent $700 million in R&D on IFR technology. It would take five years and cost $345 million to terminate the program, while it would cost $100 million to complete it over five years (1993).

Even a pro-IFR industry group, the American Nuclear Society, stated in a 1995 report that "if breeders are deployed in the future, breeding should take place only when needed and should be matched closely to the plutonium requirements of newly built reactors." (Protection and Management of Plutonium, 1995, p. 14) Of course, any 'plutonium requirements' could easily be filled with existing weapon-grade and separated plutonium.

In sum, the overall prospects for breeder reactors are very poor. As of 1996, the IFR remains canceled. Superphoenix in France has been shut down since July, 1990, and is unlikely to reopen. Only Japan retains interest in these plants - even though its reactor, MONJU, remains shut down and the Japanese nuclear program is undergoing a reassessment (Berkhout, 1993, p. 10). Chow and Solomon project that fast reactors will not be profitable until yellowcake price reaches $220/LB - an event that is not expected for about 100 years (Chow and Solomon, 1993, p. xvii).

Estimated costs of various disposition options

Cost estimates by US contractors and others on the ALMR project were reviewed by the NSC committee and found to be of uncertain validity. An independent study by the Electric Power Research Institute (EPRI) is probably more credible and this study has been used as the foundation for most independent costing in these types of reactors. The costs this study generated are shown in Table 7.2.

101

Based on the data in Table 7.2, for an advanced LWR the calculated 30-year cost in 1996 dollars is $4.3 billion of which $1.6 billion is capital cost, $2.3 billion is fixed operating cost, and $.4 billion is incremental operating cost at 80% capacity. Similarly, for a 1 GWe ALMR, the 30-year cost in 1996 dollars is about $1.3 billion more, with a total cost of $5.6 billion of which $2.2 billion is capital costs, $2.9 billion is fixed operating costs, and $0.5 billion would be required for incremental operating costs (Nuclear Wastes: Technologies for Separations and Transmutation, 1996, p. 78).

Table 7.2
Generic capital, operations and maintenance costs for various reactor types - 1996 dollars

Advanced Reactor Type	Overnight Capital Costs (per rated kW)	Ops & Maint. Costs Fixed[a] ($/kWe/yr.)	Incremental[b] (cents/kWh)
Large Evolutionary LWRs	$1,650	77.1	0.14
Mid-sized Passive LWRs	$1,870	91.2	0.14
Liquid Metal Reactors	$2,200	94.5	0.19

[a]These operating costs are essentially independent of actual capacity factor, number of hours of operation, or of kilowatts produced. They include labor charges for plant staff.
[b]These variable operating costs and consumables are directly proportional to the amount of kilowatts produced. They include chemicals consumed during plant operations.

Sources: Electric Power Research Institute (EPRI) (1989), *Technical Assessment Guide, Electricity Supply - 1989*, Vol. 1, Rev. 6, Special Report EPRI P-6587-L, Palo Alto, California, and National Research Council, *Nuclear Power: Technical and Institutional Options for the Future* (1992), Committee on Future Power Development, National Academy Press, Washington, DC.

After reviewing the costs of the various reactor options for disposing of plutonium through reactor burning, the National Academy of Sciences calculated what costs these options would accrue to dispose of 50 metric tons of weapon grade plutonium. The costs shown in Table 7.3 are the result of updating these computations to 1996 dollars, but they do not include the costs of the 'additional requirements' facilities. MOX facilities are an additional requirement for almost every disposal option, and the cost of these facilities have been calculated in Chapter 6 of this book. In addition, taking waste storage costs into account further raises the cost of burning the plutonium in LWRs by $4350 to $10,900/kg in 1996 dollars and this would have to be added to the costs of disposition shown in Table 7.3 (Chow and Solomon, 1993, p. xvii).

Table 7.3
Approximate costs to dispose of 50 metric tons of plutonium - 1996 dollars

Options	Cost To Dispose Of 50 MT Pu	Additional Requirements
MHTGCR/No reprocessing	$6.2 billion	
ALWR/No reprocessing	$3.4-$5.8 billion	MOX facilities
Complete WPPSS	$2.1 billion	MOX facilities
Existing LWR's	$320 million	MOX facilities
Existing Reactors (Foreign)	$320 million or more	MOX facilities
Existing Reactors (Canada)	$320 million or more	MOX facilities
ALMR	$5.9 billion	New processing technologies

Source: National Academy of Sciences (1994), *Management and Disposition of Excess Weapons Plutonium (Pre-Publication Copy)*, Committee on International Security and Arms Control, National Academy of Sciences, National Academy Press, Washington, DC.

Currently operating LWRs could be modified for plutonium disposition in 8 to 10 years at a cost of about $50-100 million per year in 1996 dollars, exclusive of the cost of construction of any test facilities. Use of an ALMR/IFR in this role may prove to be more economical than aqueous reprocessing, but this has not been demonstrated. Further, light water reactors specially designed for burning plutonium would consume plutonium at a rate similar to an ALMR with a 0.65 breeding ratio. Compared to a LWR, an ALMR/IFR is more expensive and would take 15 to 20 years to develop and demonstrate. All this would cost roughly twice the annual cost of a LWR (Nuclear Wastes: Technologies for Separations and Transmutation, 1996, p. 5). Thus, in spite of its technical feasibility, the National Academy of Sciences noted in its 1994 report that:

.... in the current nuclear fuel market, the use of plutonium fuels is generally more expensive than the use of widely available LEU fuels - even if the plutonium itself is 'free' - because of the high fabrication costs resulting from plutonium's radiological toxicity and from the security precautions required when handling it. As a result, while most of the world's roughly 400 nuclear reactors could in principle burn plutonium in fuel containing a mixture of uranium and plutonium (mixed-oxide or MOX fuel), few - and none in the

United States - are currently licensed to do so (Management and Disposition of Excess Weapons Plutonium, 1994, p. 5).

Many other options for plutonium disposition are available, and these will be discussed at length in the chapter that follows. Like the use of breeder reactors, a number of these options may also prove to be prohibitively expensive.

Specific experience with plutonium fuel use

The economic realities of commercial breeder reactors and MOX burning in light water reactors have been faced by the US, Japan, Germany, Great Britain, Belgium, Russia, and France over the last twenty years, and the result has been basically the same in all six countries. The following sections give the specific experiences of each country.

France

The French have 16 MOX-burning reactors. These pressurized water reactors (PWRs) are loaded with one-third of the core containing MOX fuel composed of 6% plutonium and 94% depleted uranium. Under normal operation, the 16 French PWRs need 130 tons of MOX each year. This MOX contains about 8 tons of plutonium. Each year, the same amount of irradiated MOX is taken out of the reactors and this spent fuel contains about 6 tons of plutonium of "lesser value" because the share of nonfissile isotopes of plutonium goes up with irradiation. Thus actual plutonium consumption is about 25% (Vendreyes, 1992, pp. 61-64). The fuel division of the French nuclear agency has estimated that the use of MOX instead of LEU in France's 16 reactors during the period 1995 to 2000 would result in excess costs of about $400 million ($1996) (Schneider, 1996, p. 4).

French Fast Neutron Reactors can burn plutonium of any isotopic composition. When roughly 20% plutonium and 80% depleted uranium are used in the core, the amount of plutonium produced will be larger by roughly 20% than the amount of plutonium consumed by fission. The French claim they can design these reactors so the amount of plutonium they produce can be regulated at will. A fast reactor with a full core of plutonium should consume about 700 Kg of Plutonium per year in a 1000 MWe reactor (Vendreyes, 1992, pp. 61-64).

Actual French experience in this area has confirmed that fast breeder reactors are uneconomical to operate. The Phoenix, a prototype fast breeder reactor started operation in 1973 (Takagi, 6/22-27/1996, p. 1). Superphoenix, the first commercial-scale fast breeder reactor, has had continuous problems since it began operations in 1986. Leaks in the liquid sodium coolant system have persisted and the reactor was shut down in 1990 after it had operated only 174 days in eight years. France's Atomic Energy Commission (CEA) claims the Superphoenix could burn 200 kg of plutonium a year, and in March, 1994, it released plans to make modifications to achieve that goal. While the CEA has not said what these modifications will cost, the French government admits it has abandoned the idea that the Superphoenix will ever make money (Rothstein, 1994, pp. 8-9). Officially, the total investment in the Superphoenix FBR was put at $6.4 billion in 1996. Considering only these investment costs, and excluding all operations and maintenance, electricity generated at the Superphoenix facility costs over $1.15 per kWh (Schneider, 1996, p. 3).

As of 1996, COGEMA was offering MOX fuel fabrication contracts for the post-2000 period at prices pegged to the price of LEU fuel, a practice that leads to MOX prices that are well below cost. A confidential German utility document stated in 1995 that "in the case of reprocessing, COGEMA offers with it the MOX-fuel element fabrication for a price, which if possible shall not exceed 1.25-times the equivalent uranium dioxide-fuel element fabrication price." (Schneider, 1996, p. 4) Further, an analysis of COGEMA's contracts for the period 2000 and later shows that they contain provisions to store spent fuel for the contract holders for at least ten years with an option for the customer to recover its spent fuel, unprocessed, after that time and return it to its own country. Such a contract is primarily for long-term storage, not for MOX - a situation that complicates an analysis of the views of potential MOX customers regarding the actual use of MOX fuel (Schneider, 1996, p. 7).

The French civilian power industry owes bondholders billions of francs and the French government no longer guarantees nuclear bonds. An independent study by the International Project for Sustainable Energy Paths claims that the French nuclear power industry is "the world's most indebted corporation" and cites continual overestimation of demand and underestimation of costs. Official estimates of costs by the French government were based on increases of 1.5% per year, but costs actually rose at 5-6% per year. Overall, government nuclear power costs have probably been underestimated by at least 60% (Rothstein, 1994, pp. 8-9).

Germany

Plutonium reprocessing in Germany dates back to 1966. In 1970, MOX fuel was loaded in the reactor at Obrigheim. The Alkem 25 tHM/yr. reprocessing plant was opened in 1965 and then closed by Siemens in 1991. A new 129 tHM/yr. MOX plant was built next to the Alkem plant, but was abandoned in 1995. Presently, 21 German LWR's produce about 450 tons of spent fuel per year which contain about 4.5 tons of plutonium. So far, 18 tons of plutonium have been separated, of which 7 tons were fabricated into MOX and the rest stored in different forms. At present, 11 German LWR's and 2 German boiling water reactors (BWRs) are licensed to use MOX. A total of 250 MOX assemblies have been loaded into these reactors (Decressin et al., 1996, p. 4).

Construction of Germany's Kalkar breeder reactor was halted in 1986 before the reactor became operational. The entire project was abandoned in 1991 after total expenditures of over $4.6 billion ($1996). In 1995 the 40 acre reactor site was sold to an amusement park developer from the Netherlands who plans to market it as the "Nuclear Water Wonderland." (A Great Fixer-Upper, 1996, p. 8)

Belgium

Of the 400 tons of MOX manufactured by the end of 1995, 270 tons (equivalent to 13 tons of plutonium) were made in Belgium. MOX has been used in the pressurized water reactor at the Belgian Nuclear Research Center at Mol since 1963. The 35 tHM/yr. MOX fuel fabrication plant at Dressel has been operated since 1973. Belgonucleaire fabricated more than 180,00 fuel rods between 1977 and 1985, and 150,000 rods between 1986 and 1995 for a large number of clients and in 1995, the first commercial power reactors in Belgium began using MOX fuel (Decressin et al., 1996, pp. 2-3).

United Kingdom

Reprocessing started at Sellafield in the 1950's. Civilian plutonium was incorporated in British Nuclear Fuel Ltd.'s (BNFL) operations after 1964. About 20 tons of fast reactor MOX fuel has been made since 1970. In 1996, BNFL had an 80 ton stockpile of reactor grade plutonium, half in irradiated fuel in storage ponds, and half as separated plutonium dioxide in storage (Decressin et al., 1996, pp. 4-5).

Canada

Canada's CANDU (deuterium-uranium) reactor has a design that allows it to handle full mixed-oxide cores, and unlike light water reactors, the CANDU does not have to be shut down for refueling. In addition, the fissile content of the fuel can be burned down to a lower level than in a light water reactor so more energy is extracted per kilogram of plutonium. By one estimate, two standard design CANDU reactors could transform 50 metric tons of weapons plutonium into spent fuel in 25 years (Charnetski and Rauf, 1994, p. 23-24).

Japan

The Japanese program is now about 30 years old. Early forecasts about how cheap this reprocessing/breeder program would be have been proved wrong - Japanese breeder reactors are now estimated by some sources to be 5 to 15 times more expensive to run than conventional nuclear power plants (Sanger, 1994).

At the time of the plutonium shipment from France to Japan on the Akatsuki-maru (1992-1993), the Japanese claimed the shipment was urgent because the plutonium was needed to fuel MONJU (Takagi, 1/10/1996). Japan planned to obtain 50 tons of plutonium by 2011, mostly from European plants, and it planned to open a new reprocessing plant of its own (Wald and Gordon, 1994). However, Japan has delayed the reprocessing plant and several breeder reactors from 2010 to

2030. These delays were partly due to citizen pressure and partly due to sharply increasing financial risks. On June 7, 1996, the Japanese Ministry of International Trade and Industry announced a moratorium on the construction of a follow-on breeder reactor and a re-evaluation of the Nuclear Energy Development Plan, stating that "going forward with plans will be difficult until the understanding of the people of the nation has been gained." (Takagi, 6/22-27/1996, p. 1)

The MONJU breeder reactor was activated in March, 1994, at a cost of $5.3 billion in 1996 dollars. The reactor took 20 years longer to reach operations than had been forecast in 1967 by the Nuclear Energy Long Term program. The December, 1995, sodium leak at MONJU was the largest ever to occur in the piping of an operating reactor and was also the worst in terms of the leakage rate. Even the most optimistic observers believe it would take three to four years for MONJU to restart and there is virtually no prospect for construction of the 600 MW demonstration fast breeder (DFBR) that is to follow MONJU (Takagi, 1/10/1996).

Russia

In the Former Soviet Union (FSU), the breeder reactor was originally planned as a source of plutonium for bombs. Fast breeder reactor research began in the FSU in the 1950's. In the 1960's, Russian pilot programs such as BR-5, IBR-2, IBR-30 were instituted on variants of breeder reactors. Tests have been completed in Obninsk at the Institute of Energy and Physics on two partial cores in the fast breeder BR-10 using plutonium dioxide manufactured from weapon-grade plutonium. There are also plans to construct 3 or 4 fast breeder reactors in the Urals based on the BN-800 design (Kushnikov, 1995, pp. 28-29).

Implementation of the closed fuel cycle started with the MAYAK facility at Chelyabinsk, a multi-purpose facility for reprocessing spent fuel from VVER 440 reactors, from BN 350 and BN 600 fast breeder reactors, and spent fuel from ice breakers, submarines, and research reactors. The capacity of this plant is 400 tons/year. By December, 1995, this plant had reprocessed 3000 tons of spent fuel. A new Russian plant with a capacity of 1500 tons/year is scheduled to begin operations in 2005 in the Krasnoyarsk region (Kushnikov, 1995, pp. 26).

The 1992 G-7 conference decided to upgrade Russia's reactors to avoid further safety problems. By 1996, nearly $900 million had been spent to correct these problems, and an October, 1995, report by the Center for Strategic and International Studies called for an additional 10 year, $20-30 Billion effort to solve Russia's other reactor problems. DOE sought $76 million for this purpose in 1995 but only received $30 million (Capping Fallout from Russia's Nuclear Legacy, 1995). Russia could replace all its reactors with natural gas and coal- fired power plants for $6-$7 billion over six or seven years. In contrast, the International Atomic Energy Agency has said that upgrading existing Russian reactors to Western safety standards would take from $28-$127 billion in 1996 dollars (Wald and Gordon, 1994). Meanwhile, export earnings by MINATOM were $1.2 Billion in 1994 and $1.5 Billion in 1995, and MINATOM has contracted with Iran to sell one reactor for

$800 million and build two more for about $1 billion (Ingwerson, 1995). Russia envisions keeping weapons plutonium in storage for the next few decades and using its surplus of reactor-grade plutonium for MOX fuel because any delays in reprocessing reactor-grade plutonium will allow it to become more radioactive and thus, harder to handle (Wald and Gordon, 1994).

Conclusion: specific cost additions arising from the use of plutonium in commercial reactors

The economics of plutonium burning have been repeatedly investigated and rejected. Chow and Solomon looked at five options for using plutonium in reactors. The options and their costs, in 1996 dollars, are:

1. Use plutonium as fuel in existing fast reactors without reprocessing. Using weapon-grade plutonium in this manner would cost $19,500/kg.
2. Use LWR's with one-third or partial MOX fuel without reprocessing. The cost for this is $8,300/kg with weapon-grade plutonium.
3. Use LWR's with full MOX fuel loads without reprocessing. The cost for this is $6,000/kg with weapon-grade plutonium.
4. Store plutonium for 20 or more years. Cost: $4,100/kg.
5. Mix plutonium with waste and dispose of it as waste. Cost: $1,080/kg in marginal costs over storing the waste alone - which would lead to costs of about $5,200/kg (Chow and Solomon, 1993, pp. xxi, xxii).

None of these options has any commercial value. In the first three, the extra costs of handling plutonium because of its radioactivity, toxicity, and potential weapon use outweigh any benefits. Further, storage sites will not be ready until 2010 at the earliest, and when storage costs are taken into account, they raise the cost of burning plutonium in LWRs by $4400 to $10,900/kg.

The National Academy of Sciences (NAS) study estimates that a new MOX fabrication facility would cost between $425 million and $1.3 billion in 1996 dollars and would take about a decade to complete (Management and Disposition of Excess Weapons Plutonium, 1994, pp. 159-160). The cost of MOX fuel fabrication is estimated to cost over $2100 per kilogram of heavy metal, about six times the fabrication cost of LEU fuel (Nuclear Fuel, 1996, p. 2). In addition, it is not clear that the NAS study included the additional costs of security and handling that would be certain to accompany any plutonium processing or storage.

Cost estimates for geologic repository disposal of spent fuel from commercial power reactors are about $400,000 per ton of heavy metal in 1996 dollars. However, the cost to dispose of a ton of plutonium would be higher because it must be diluted to make re-extraction difficult. Assuming a cost on the order of several million dollars per metric ton of plutonium, 1996 total disposal costs would range from \$100-\$310 million for 50 metric tons of plutonium (Makhijani and Makhijani, 1995, p. 66).

Because of these significant cost additions, the use of plutonium in civilian reactors has no economic rationale and is accompanied by a large proliferation risk. Chow and Solomon estimated that MOX fuel use will not be feasible for 50 years until the price of uranium-bearing yellowcake reaches \$110/LB. They further projected that fast reactors will not be profitable until yellowcake price reaches \$240/LB in about 100 years (Chow and Solomon, 1993, p. xvii). Of course, if either of these conditions occurred, it is likely that neither uranium nor plutonium would be a competitive source of energy. Thus, plutonium burning, which is not attractive economically, requires complex fuel fabrication, and is feasible for one or two cycles only is an unlikely solution for the disposition problem (Lombardi, 1994).

References

'A Great Fixer-Upper' (1966), *The Bulletin of the Atomic Scientists*, Vol. 52, No. 2, p. 8.

Berkhout, Franz (1993), *Fuel Reprocessing At THORP: Profitability and Public Liabilities*, Center for Energy and Environmental Studies, Princeton University, Princeton, NJ, p. 10.

Bradley, Carol (1993), 'Crapo Asks To Fund More Work On Reactor', *The Idaho Statesman*, June 10.

'Capping Fallout from Russia's Nuclear Legacy' (1995), *The Christian Science Monitor*, November 8.

Charnetski, Joanne and Rauf, Tariq (1994), 'Let Canada Cremate Nuclear Swords', *Defense News*, October 3-9, pp. 23-24.

Chow, Brian G. and Solomon, Kenneth A. (1993), *Limiting the Spread of Weapon-Usable Fissile Materials*, National Defense Research Institute, RAND, Santa Monica, CA, 1993, pp. xvii-xxii.

Decressin, A., Gambier, D.J., Lehmann, J., Nietzold, D.E. (1996), *Experience and Activities in the Field of Plutonium Recycling in civilian Nuclear Power Plants in the European Union*, Paper Presented at the International Conference on Military Conversion and Science: Utilization/Disposal of the Excess Weapon Plutonium: Scientific, Technological and Socio-Economic Aspects, Como, Italy, pp. 2-7.

Ingwerson, Marshall (1995), 'Marketing Nuclear Plants For an Energy-Hungry World', *The Christian Science Monitor*, November 8.

Integral Fast Reactor (1993), Department of Energy, Washington, DC.

Johnston, J. Bennett (1993), 'IFR For A Safe World', *Letter to the Washington Post*, October 13.

Kushnikov, Viktor (1995), 'Radioactive Waste Management and Plutonium Recovery Within the Context of the Development of Nuclear Energy in Russia', *Final Proceedings: US Department of Energy Plutonium Stabilization and Immobilization Workshop*, pp. 26-29.

Lindsay, Richard W. (1994), 'Reactor Solves Problems As It Offers Benefits', *The Idaho Statesman*, Boise, Idaho, June 26.

Lippman, Thomas W. (1993), 'Disputed Nuclear Program Reborn', *The Washington Post*, April 13.

Lombardi, Carlo (1994), *Weapon-grade Plutonium, Annihilation Via Thermal Fission in Unconventional Non-fertile Matrices*, International Congress on Conversion of Nuclear Weapons and underdevelopment: Effective Projects from Italy, Rome.

Makhijani, Arjun and Makhijani, Annie (1995), *Fissile Materials In A Glass, Darkly*, IEER Press, Takoma Park, Maryland, pp. 26-27, 66.

Management and Disposition of Excess Weapons Plutonium (1994), Committee on International Security and Arms Control, National Academy of Sciences, National Academy Press, Washington, DC, pp. 5, 159-160.

Nuclear Fuel (1996), September 9, 1996, p. 2.

Nuclear Wastes: Technologies for Separations and Transmutation (1996), Committee on Separations Technology and Transmutation Systems, Board on Radioactive Waste Management, Commission on Geosciences, Environment, and Resources, National Research Council, National Academy Press, Washington, DC, pp. 5, 78.

Protection and Management of Plutonium (1995), American Nuclear Society Special Report, pp. 10, 14.

Rothstein, Linda (1994), 'French Nuclear Power Loses its Punch', *The Bulletin of the Atomic Scientists*, July-August, pp. 8,9.

Sanger, David E. (1994), 'Japan, Bowing to Pressure Defers Plutonium Projects', *The New York Times*, February 22.

Schneider, Mycle (1996), *Plutonium Fuels*, World Information Service on Energy, Krasnoyarsk, Russia, pp. 3-7.

Schulze, Joachim (1992), 'Burning of Plutonium in Light Water Reactors (MOX Fuel Elements) Compared To Other Treatment', *Working Papers of the International Symposium on Conversion of Nuclear Warheads for Peaceful Purposes*, Rome, Italy, pp. 65-74.

Takagi, Dr. Jinzaburo (1/10/1996), Citizens' Nuclear Information Center, 1-59-14-302 Higashi-nakano, Nakano-ku, Tokyo 164, Japan.

Takagi, Jinzaburo (6/22-27/1996), *Japan's Plutonium Program and Its Problems*, Third International Radioecologica Conference on The Fate of Spent Nuclear Fuel and Reality, Krasnoyarsk, Russia, pp. 1, 9.

Takagi, Jinzaburo (1995), 'Sodium Leak Hits Achilles' Heel of FBR MONJU', Citizens' Nuclear Information Center, Tokyo, December 9.

Uematsu, Kunihiko (1992), 'The Technological and Economic Aspects of Plutonium Utilization in Fission Reactors', *Working Papers of the International Symposium on Conversion of Nuclear Warheads for Peaceful Purposes*, Rome, Italy, pp. 109-115.

Vargas, Dale (1993), 'Scientists tout new nuke reactor as safe', *Sacramento Bee*, May 17.

Vendreyes, Georges (1992), 'Plutonium Burning In Fast Reactor and As MOX Fuel', *Working Papers of the International Symposium on Conversion of Nuclear Warheads for Peaceful Purposes*, Rome, Italy, pp. 61-64.

Vendreyes, Georges (1994), *Energy for Mankind and the Use of Fissile Material from Disarmament in Nuclear Reactors*, International Congress on Conversion of Nuclear Weapons and Underdevelopment: Effective Projects from Italy, Rome.

Wald, Matthew L. and Gordon, Michael R. (1994), 'Russia And US Have Different Ideas About Dealing With Surplus Plutonium', *NY Times News Service*, August 19.

8 A comparison of proposals for disposition of plutonium from warheads

Introduction

Virtually everyone now recognizes the dangers posed by surplus plutonium stocks, a problem that has spawned volumes of research on alternative means of plutonium disposition. This chapter provides an overview of the most frequently mentioned alternatives for dealing with surplus weapon-grade plutonium and its most troubling problem: a level of radioactivity that is too low to deter committed terrorist or subnational groups (groups acting without government authorization) from reusing it in weapons.

Both weapon-grade and reactor-grade plutonium have additional properties that make safe, secure storage an absolute requirement:

1. Reactor grade plutonium with any level of radiation is a potentially explosive material.
2. Developing an effective, simple bomb design is not appreciably harder with reactor-grade plutonium than it is with weapons-grade plutonium.
3. The hazards of handling reactor-grade plutonium, although somewhat greater than those associated with weapons-grade, are of a type that can be met by applying the same precautions.
4. The need to protect against diversion is the same for all grades of plutonium (Garwin, 1993).

Long-term plutonium disposition falls into three general categories of options:

1. Indefinite storage.
2. Minimized accessibility using physical, chemical or radiological barriers to reduce access.
3. Elimination by making plutonium completely inaccessible through complete burning or by launching it into deep space (Management and Disposition of Excess Weapons Plutonium, 1994, p. 12).

Whatever approach is chosen, a generally accepted final objective is to make surplus weapon-grade plutonium at least as inaccessible as the plutonium contained in spent reactor fuel. There is so much spent reactor fuel world-wide that if this objective is met, any terrorist or subnational group would use plutonium in spent fuel instead of less accessible surplus weapon plutonium (Management and Disposition of Excess Weapons Plutonium, 1994, pp. 12-13).

In addition, any plutonium disposition method must be socially responsible, guarding the health and safety of workers and local residents alike. With this in mind, Makhijani and Makhijani have suggested the following criteria for evaluating the acceptability of the various options for plutonium disposition:

1.   Security aspects. The treatment, storage and disposal of plutonium must make it as difficult to re-extract plutonium as it is to separate it from reactor fuel.
2.   Timeliness. The option should be available as soon as possible.
3.   Risk of accident. The risk of catastrophic accidents must be minimized.
4.   Health, environmental protection and safety. The option should comply with all applicable regulations, including the reality of increased handling required by some options.
5.   Production. The option should not encourage further plutonium production (Makhijani and Makhijani, 1995, pp. 19-20).
6.   Cost. Many of the arguments for using plutonium are cost-driven. In these cases the arguments against using plutonium must also be based on cost.

Disposition forms

Highly enriched uranium (HEU)

The ease with which highly enriched uranium can be rendered relatively harmless for weapon use through downblending and the demand for downblended HEU as reactor fuel both indicate that little discussion of alternative disposition options for HEU is necessary - if its use as reactor fuel can be accomplished while the nuclear industry is still sufficiently viable to consume the fuel. However, the amounts of materials involved during and after downblending do create storage and monitoring problems. Each kilogram of HEU must be diluted with 70 kilograms of depleted uranium (DU) to make .9% low enriched uranium reactor fuel, so disposition of 200 tons of HEU would require 14,000 tons of DU and would generate 14,200 tons of reactor fuel that must be stored and monitored.

If a decline in the nuclear power industry destroys the demand for downblended HEU and the remaining material must be treated as a waste, disposal costs have been calculated at $20 per cubic foot in 1996 dollars. Estimates are that each waste canister for uranium could hold 15 kilograms of material and that a 55 gallon drum could hold six canisters. If these figures are correct, disposition of 200 tons of HEU

113

would generate 16,700 tons of low-level waste that would have to be stored in 186,000 drums. This would consume 1.3 million cubic feet of disposal space at a cost of about $26 million (Campbell and Snider, 1996, p. 5). This figure does not include downblending costs or any other expense required to prepare HEU for disposition.

Plutonium

There is considerable debate about the optimal form in which to store plutonium. For the near-term, the 1994 National Academy of Sciences report recommended one of two options:

1. Storage as intact weapon components (pits). This postpones the costs of conversion and the Academy felt that intact pits do not pose any greater danger than deformed pits or metal ingots. This option is currently being exercised at the Pantex Plant in Amarillo, Texas, where nuclear weapons are dismantled.
2. Deformation of pits, if it can be done at low cost and acceptable environmental, safety and health risks (Management and Disposition of Excess Weapons Plutonium, 1994, pp. 10-11). Deformation would make the pits unusable in weapons unless they were reshaped.

Leaving surplus plutonium in the metallic form it had in warheads causes proliferation concerns because of the ease with which plutonium metal can be converted back into military-usable shapes. As an additional problem, plutonium oxidizes under normal atmospheric conditions and the US and Russia have little experience in dealing with oxidation in stored plutonium (Uematsu, 1992, pp. 109-115). For example, a survey of 30 representative plutonium storage cans at Los Alamos National Laboratory showed that while none had failed, one-third of the containers had signs of deterioration that could lead to failure if not properly stabilized and repackaged (Leslie, 1994, pp. 2-3).

Some have proposed converting plutonium to plutonium dioxide as a partial solution to the oxidation problem. This would increase storage time and make it difficult to convert back to military uses. The nuclear nations do have considerable experience in storing large amounts of PuO_2 and there is an international consensus by the IAEA on storage methods (Uematsu, 1992, pp. 109-115). However, converting to PuO_2 would require new plants, new processes, and a substantial capital investment - an industrial scale plant to convert plutonium pits to oxide might cost $100 million or more (Bunn, 1996, pp. 4-5). In addition, this process would undoubtedly produce new waste streams. The cost of either aqueous reprocessing or a plutonium oxide plant would have to be incurred if the MOX option is selected because the gallium used as an alloying element in plutonium pits must be removed before MOX can be fabricated (Toevs and Beard, 1996). The DOE estimated that if this option is pursued it would add about $200 million to the cost of MOX fabrication (Technical Summary Report For Surplus Weapons-Usable Plutonium Disposition, 1996, pp. 6-3).

Direct disposal of plutonium through burial might lower the risks of radiation, but since the plutonium could conceivably be retrieved, many researchers have suggested that one must 'denature' weapon-grade plutonium to prohibit reuse. If weapon-grade plutonium was denatured with Pu-238, this material would have to be produced in an accelerator - an expensive and time-consuming project. Using reactor-grade plutonium is also expensive and does not guard against reuse of the weapon-grade material. 'Spiking' weapon-grade material with Co60 to make it hard to handle only lasts a relatively short time because of the short half-life of Co-60 (Protection and Management of Plutonium, 1995, p. 10).

Based on these drawbacks, there is general agreement among most researchers that in the short- to medium-term, plutonium should simply be mixed with wastes to raise its radioactivity to the 'spent fuel standard' as suggested by the National Academy of Sciences in 1994 (Management and Disposition of Excess Weapons Plutonium, 1994, pp. 10-11).[1] A 1995 report by the American Nuclear Society claims that spent fuel is too radioactive to be a credible target for subnational theft or seizure, and it recommends that all plutonium be converted to the spent fuel standard either by mixing or by reactor irradiation (Protection and Management of Plutonium, 1995, p. 10).

Proponents of the spent fuel standard generally assume, as DOE's Plutonium Disposition Study did, that a surface radiation level of 100 rem would render spent fuel "inaccessible to terrorists." However 100 rem is not immediately lethal. Fanatics or terminally ill people still might approach such elements in a bare hands manner (Hardung, 1994, pp. 22, 24). Even so, the spent fuel standard appears to be appropriate in the short- and medium-run if the concerns are mainly technical ones of re-extraction and protection from diversion. It should be noted, however, that the spent fuel standard is currently unacceptable to countries that reprocess civilian fuel because it does not allow recovery of their investment in plutonium extraction (Makhijani and Makhijani, 1995, p. 4).

Beyond this, there is little agreement on plutonium disposition methods, although a consensus seems to be growing for some type of vitrification option. This chapter separates disposition options into two groups: those that appear to be non-viable for either economic or technical reasons and those that appear to be viable for both economic and technical reasons. Note that the standard for viability is higher - a viable option must be both economically and technically feasible.

115

Disposition options that appear to be non-viable for technical or economic reasons

Underground fissioning by nuclear explosion

The Russians have proposed destroying 5000 warheads with a single explosion of a 100-kiloton warhead one kilometer underground. The 20,000 kg of plutonium would be dispersed through 30,000 tons of glass melt in the cavity produced by the explosion (von Hipple, 1992, pp. 119-128). This option would create the possibility of contaminated water supplies - and it would incur all the costs associated with this contamination.

A US proposal would use small shafts to destroy five warheads at a time. This would require about 3000 detonations, depending on the stockpile level when the disposition started. However, even if one could destroy fifty warheads at a time, 300 detonations would be required, an amount equal to almost a half of the 730 US underground tests conducted to date (Plutonium - Deadly Gold of the Nuclear Age, 1992, pp. 130-138). In addition, if this option was exercised at the Nevada Test Site, it would occur directly over the area's aquifer.

The technical aspects of this option are fairly well known. Sixty-three US underground tests involved more than one explosive device, and one test used six devices. Direct costs of this option would be similar to costs of conducting underground tests of nuclear weapons. These costs ranged from $20 million to $60 million per test (Known Nuclear Tests Worldwide, 1945-1994, 1995, p. 70). Costs would be lowered to the extent that extensive telemetry and experiment design required for successful underground tests would not be required. A 2.5 meter hole 300 meters deep can be constructed at the Nevada Test Site for about $1.1 million ($1996). If 100 pits were placed in this hole, the cost per pit destroyed would be $11,000 based on the cost of the hole alone. For 20,000 pits, this would lead to a cost of $220 million for holes and a likely total cost of about $2.2 billion for the total program less costs of development and licensing (Management and Disposition of Excess Weapons Plutonium, 1994, p. 285).

These disposition costs increase significantly when the underground pollution of aquifers and other resources are considered. Speaking in 1994 at the Planetary Emergency Seminar, Prof. Zichichi, a principal organizer of the conference, said that "If we want to proceed with the dismantling and disposal of our Cold War assets in an environmentally acceptable way, then the cost will have to be on the same order as that required to build all the gadgets." (Tigner, 1994, p. 10) According to a 1996 study completed at the Brookings Institution, the cost to develop and build all the nuclear weapons in the US arsenal was about $375 billion (Schwartz, 1996).

Burning plutonium in light water reactors (LWRs)

This option requires MOX technologies to dispose of surplus weapon-grade plutonium. Reprocessing must be employed both when the plutonium is contained

116

in spent fuel and when weapon-grade plutonium - which contains gallium as an alloying agent - is used. As Chapters five, six and seven have shown, MOX burning is technically viable for plutonium from spent fuel, and it may become technically possible for weapon-grade plutonium, but it cannot be economically justified in either case in a commercial power environment. Further, the waste streams involved in the necessary reprocessing would create significant additional costs.

Burning plutonium in fast reactors

This option also requires the same technologies as (2) above, although fuel enrichment can be raised to 15-30% from the 3-4% used in a LWR. Again, as Chapters five and seven has demonstrated, this may or may not be technically viable, but it is not economically feasible (Schulze, 1992, pp. 65-74).

Burning plutonium in unconventional matrices

One additional burning option has surfaced as the drawbacks of light water and fast reactor use have become apparent. This is the option of burning plutonium in unconventional matrices as proposed by members of the Nuclear Engineering Department of the Centro Studi Nucleari Enrico Fermi in Italy.

This concept uses current PWRs with a partial core load of non-fertile oxide-type fuel with the rest of the core containing standard U-235 enriched fuel. The unconventional fuel is PuO_2 diluted in an inert matrix which is highly radiation resistant and scarcely neutron absorbent. The inert matrix may be ZrO_2, Al_2O_3, MgO or their appropriate combinations (Lombardi and Mazzola, 1994). With the core uniformly loaded with all fuel assemblies containing a suitable number of inert matrix fuel rods, no matter whether weapon-grade or reactor-grade plutonium is used, results show more than 98% of Pu-239 is fissioned. The percent of all plutonium fissioned is also quite high: about 90% of reactor grade and 93% of weapon grade plutonium (Lombardi et al., 1996, pp. 5-6). The discharged plutonium would meet the mixed waste standard (Lombardi and Mazzola, 1994).

One of the authors of this proposal has noted that the economic implications of this kind of burning are not markedly different from those for burning MOX. In addition, this method of dispositon will encounter the same problems with gallium removal as other MOX options. As a result, this method is not economically feasible. However, it appears to be preferable to MOX use because the discharged fuel would not undergo reprocessing or chemical treatments, leading to only a one-time pass-through of the plutonium. Inert matrices have not undergone systematic testing and the development of these fuels still remains to be accomplished, leaving the technical viability of this option uncertain (Lombardi, 1994).

A 1982 NASA study estimated the cost of this option at \$336,000 (\$1996) per kilogram of plutonium. These costs were increased by the assumption that the plutonium was part of a mixture of spent reactor fuel in which a ton of spent fuel would include only 17 kilograms of plutonium. Several hundred kilograms of this material could be handled at a time. This option is unlikely to be feasible due to public fears about the potential for a crash and resulting dispersion of plutonium from one of the rockets (Plutonium - Deadly Gold of the Nuclear Age, 1992, pp. 130-138).

A 1996 paper by Taylor laid out a fairly complete program for solar disposition with much lower costs than those in the 1984 NASA study. To achieve these reduced costs, the plutonium was separated from spent reactor fuel (the costs of this part of the option were not included, nor was a consideration of the waste streams this would generate) so only the plutonium could be transported to the sun. Under this scenario, the costs of getting one kilogram of plutonium into space were about \$30,000 (\$1996) and the space-related costs of disposing of 250 tons of weapon grade plutonium were estimated at about \$29 billion (Taylor, 1996, pp. 24- 25). The high degree of technical uncertainty surrounding this option is likely to substantially increase the estimated costs.

Ocean-based disposition

Three methods of ocean-based disposition have been suggested. All three methods face major obstacles both from environmental groups and from international organizations whose approval would be necessary to make such disposition legal.

Sub-seabed disposal In this option material would be placed in canisters and inserted in the 'abyssal clay formation' thirty meters below the ocean floor. This could be accomplished either through drilling or through free-falling 'penatrators' dropped from ships (Management and Disposition of Excess Weapons Plutonium, 1994, pp. 265-275). In one variant of this option, the plutonium would be packaged in small canisters with randomly designed fin patterns and then dropped in various parts of the ocean. The fin patterns would disperse the canisters and make them difficult to recover.

If ten kilograms of plutonium were placed in each canister, 10,000 canisters could dispose of the entire US and Russian stockpile of excess weapon-grade plutonium. Costs of this program have been estimated to be comparable to mined geologic disposal. While the actual emplacement of the canisters could probably be done for a few hundred million dollars, the development and demonstration needed to meet licensing requirements could cost billions of dollars (Management and Disposition of Excess Weapons Plutonium, 1994, pp. 265-275).

Concerns with this method center around long-term canister viability in salt water and the attendant chances of pollution as the plutonium migrates through the clay in which it was placed. There is also a chance of canisters being recovered from the ocean floor.

Dilution   In this option, plutonium would be converted to powder form and scattered across the oceans. If the dilution was accomplished in a uniform manner it would have an insignificant effect on the immediate levels of plutonium in the water, but it would have significant long-term effects. Shellfish, bivalves, and other ocean life are heavy metal accumulators. These organisms could re-accumulate the plutonium to levels 3000 times that of the surrounding water and insert it into the food chain (Management and Disposition of Excess Weapons Plutonium, 1994, pp. 275-279).

Since dilution could probably be accomplished for several tens of millions of dollars, the cost of this option is minimal if biological concentration is not considered. However, new standards on ocean dumping and the likely effect of this option on the food chain would both increase these costs to extremely high levels. Further, the obvious environmental concerns this option raises would make it impossible to implement ((Management and Disposition of Excess Weapons Plutonium, 1994, pp. 275-279).

Tectonic Plate Burial   This option would put plutonium in canisters that are inserted in subduction zones in the earth's crust where they would be drawn into the earth's core by tectonic plate movements. Concerns with this option center around the geologic time span required to complete disposition - the plates move at 3-10 cm per year - and the fact there is no guarantee material in subduction zones will actually subside toward the earth's core. There is also a likelihood of pollution if the canisters are breached and, of course, there is the fear that the accumulated canisters could be recovered from the disposition sites.

The costs of this option would probably be greater than those for seabed disposal. Emplacement of the canisters is likely to be more difficult and expensive because of the depth and the nature of the areas where the canisters would have to be inserted. Like seabed disposal, the development and demonstration needed to meet licensing requirements would cost billions of dollars (Management and Disposition of Excess Weapons Plutonium, 1994, pp. 265-275).

Mix and melt disposition

In a 1996 DOE report on the disposition of aluminum-clad spent nuclear fuel, one of two recommended options was "a dilution option, press and dilute or melt and dilute (as selected by further evaluation)..." (Technical Strategy for the Treatment, Packaging, and Disposal of Aluminum-Based Spent Nuclear Fuel, 1996, p. 4) Proponents of this method, such as Argonne National Laboratory, claim that reactor grade plutonium cannot easily be used to make weapons because it has 5 times the isotopic contamination of weapon grade plutonium, its radioactivity makes it more difficult to work with, and weapons made from this material would have an unpredictable yield (DeVolpi, 1995, pp. 20-21).

The mix and melt alternative proposes to melt plutonium together with spent fuel to make a mix that is too isotopically contaminated to make weapons without time-consuming and expensive processing. After this procedure is performed, the separation of isotopes would be so difficult that anyone who could do it would probably already have the option of using 'fresh' weapon-grade plutonium instead (DeVolpi, 1995, pp. 20-21). The cost of this option would be similar to the $4 billion cost of pyroprocessing proposed as part of the US Integral Fast Reactor project.

Wolfgang Panofsky, chair of the NAS Plutonium Study, notes that the mix and melt process, as a variant of pyroprocessing, is largely unproved. It would require a giant reprocessing facility capable of handling over 2000 tons of heavy metal and 50 tons of plutonium that has undergone neither engineering development nor any aspect of the regulatory process. The reactor grade plutonium in question has approximately the same isotopic concentration of Pu-239 as weapon grade plutonium. Further, no proposal for decreasing the isotopic concentration of Pu-239 makes it less desirable to terrorists since "it has been amply and convincingly documented that, while reactor-grade plutonium has not been the material of choice for past weapons builders, an explosive device with an assured yield of one to two kilotons could be built from reactor-grade plutonium by relatively elementary methods.....In the hands of a terrorist or proliferator such devices would be formidable indeed." (Panofsky, 1996, pp. 3, 59)

Transmutation

Complete elimination of plutonium is only possible through two means. First, one could wait until natural radioactive decay destroys it - this would take thousands of years. Second, one can theoretically transmute plutonium by using some technique to bombard its nuclei and split them into fission products in a particle accelerator or through nuclear reactions in a reactor (Developing Technology to Reduce Radioactive Waste May Take Decades and Be Costly, 1993, p. 11). Most elements created by transmutation would have much shorter half-lives than plutonium. Thus, the potential benefits of transmutation could be:

1. A reduced volume of material.
2. Reduced radioactive life of materials.
3. Less risk of human intrusion into storage areas (Developing Technology to Reduce Radioactive Waste May Take Decades and Be Costly, 1993, p. 4).

Waste transmutation would take many billions to develop and is not possible before 2015. DOE believes it is not economically justifiable since a waste repository would still be needed. Complete transmutation systems of the kinds shown in Table 8.1 include a reactor or accelerator to transmute reprocessed fuel, a

120

spent fuel reprocessing and waste separation facility, a fuel fabrication facility, and storage facilities for spent fuel and residual wastes (Developing Technology to Reduce Radioactive Waste May Take Decades and Be Costly, 1993, pp. 3-5).

Table 8.1
Potential transmutation technologies costs of transmutation and other non-burning or technical fixes

Potential Programs	Sponsor	Units & Time To Destroy 90% Of LWR Actinide Waste In 2010	Schedule/ Cost($1996)	Destroys Actinides	Destroys Fission Products
Advanced Liquid Metal/ Integral Fast Reactor (ALMR/IFR)	DOE, GE Argonne	19 Units 200 years	Start: 2015 Operate: 200 yr. Ops Cost: $35B $5.4B (1st reactor) $4.4B/Unit for Additional Units	Yes	No
Accelerator Transmutation Project (ATW)	LANL	19 Units 40 years	Start: 2016 Develop: $5.4B Total: $140B	Yes	Some Including Pu & U
Phoenix Accelerator	Brook- haven Natl. Lab	1 or 2 units 25 years	Development Time: 15-20 yr. Develop: $32B	Yes	Some But Not Pu or U
Particle- Bed Reactor (PBR)	Brook- haven Natl. Lab	20-70 Units 40 years 150 yr. for Pu	Development Time: 16 yr. Develop: $1.4B	Yes	Yes
Clean Use Of Reactor Program (CURE)	Hanford/ Westinghouse		Rsch: $80- $175 M	No	Yes

Source: *Developing Technology to Reduce Radioactive Waste May Take Decades and Be Costly* (1993), GAO/RCED-94-16, United States General Accounting Office, Washington, DC, December.

One transmutation technique called 'photofissioning' employs fission induced by photons from high energy gamma rays. Unfortunately, this technique shares a

121

common problem with other transmutation methodologies - it would require reprocessing and hence, is likely to be unacceptable on the basis of both proliferation and waste generation concerns (Makhijani and Makhijani, 1995, pp. 98-100). The US Government Accounting Office (GAO) has noted that where transmutation is concerned, "the reprocessing and separating of the waste are more difficult technical problems than transmuting the long-lived elements from the waste." (Developing Technology to Reduce Radioactive Waste May Take Decades and Be Costly, 1993, p. 13)

Even though waste generation is a major problem, transmutation itself creates many additional difficult problems because it is only partially researched and is relatively undeveloped. The National Research Council claims that any advanced transmutation facility would require extensive R&D to determine if it was even feasible. This research and evaluation process would take more than 20 years and would result in much higher costs. And even after a separations and transmutation (S & T) system was developed, it would still require many decades to centuries to achieve any significant net reduction in the total amount of weapon-grade materials either in or outside of spent fuel (Nuclear Wastes: Technologies for Separations and Transmutation, 1996, p. 5).

As a further complication, a separations and transmutation (S&T) system would require the integrated operations of a large number of systems. Approximately eighty LWRs, twenty transmuters, two reprocessing plants and a few fabrication plants would all have to operate as a fully integrated system. This would create many more problems than using the LWRs required for a simple, once-through system (Nuclear Wastes: Technologies for Separations and Transmutation, 1996, p. 6).

These added complication and uncertainties create significant excess costs for a separations and transmutation (S&T) disposition system compared to a once-through LWR system. Estimates of the total costs to dispose of the likely amount of surplus weapon-grade materials are uncertain but it is probable that these costs would be no less than $50 billion and they could easily exceed $100 billion in 1996 dollars. And if the transmutation system was linked to the power generating grid, the additional cost of generating wholesale electricity could increase from 2%-7%, resulting in a cost increase of about $25 to $80 billion (Nuclear Wastes: Technologies for Separations and Transmutation, 1996, p. 7).

Disposition options that appear to be viable on both economic and technical grounds

As both the preceding section and Chapters Five, Six and Seven show, none of these disposition options provides both the technical and economic benefits necessary to make it a viable alternative for plutonium disposition. However, variants of two general options - storage and irretrievable disposal - remain viable candidates for intermediate and long-term plutonium disposition. The only assumption necessary to make these options viable is the single proposition, already demonstrated in the previous chapters, that plutonium has no economic value and is best treated as a waste product.

Direct disposal in surface storage

Direct disposal places nuclear materials directly into some storage mode without additional reprocessing or burning. One option for disposing of plutonium by direct disposal is surface storage where nuclear materials could be monitored and potentially retrieved.

Management costs associated with direct disposal are calculated by the German government to be about $3600/kg ($1996). This compares favorably with a $5400/kg ($1996) management cost for fuel reprocessing (assuming post-2002 THORP prices), and fuel reprocessing management costs do not include other penalties associated with recycling plutonium (Berkhout, 1993, p. 8). For example, a penalty is applied to extended dry storage of spent fuel as the Pu-241 decays to Americium 241. MOX fuel fabricators will not take spent fuel containing more than 1.5%-2.5% Americium 241, and the cost to remove it chemically is $20/gram of plutonium (Plutonium Fuel: An Assessment, 1989, Table 14). This penalty applies only when reprocessing and MOX fabrication are contemplated.

In 1994, civilian plutonium reprocessors charged between $2 and $4 per gram to store separated plutonium. More recent information from interviews reported by Berkhout suggest plutonium storage costs of $4.35 per gram per year ($1996) (Berkhout et al., 1993, p. 169). At these prices, storing 50 metric tons of plutonium for a decade would cost from $1 to $2 billion (Management and Disposition of Excess Weapons Plutonium, 1994, pp. 126-127). This agrees with estimates by Fetter who calculated that a monitored storage facility for 50 metric tons of plutonium had an estimated capital cost of $206 million ($1996) with an operating cost of $34 million per year (Bloomster et al., 1990, pp. 12-13). [Similar storage facility operating costs were estimated by Johnson to be $30 million in 1996 dollars (1992).] Based on these figures, preliminary estimates by Fetter are considerably lower than the costs charged by civilian reprocessors. Fetter calculates that storing plutonium would only cost about $1.60 per gram per year in 1996 dollars. Thus, storing 200 metric tons would cost roughly $320 million per year for a net present value cost of about $3.2 billion with a 10% discount rate (Fetter, 1992, pp. 144-148).

Another US study done in 1990 estimated that a facility for storing 50 tons of plutonium dioxide in 12,500 canisters, each holding 4 kg of plutonium, would have a capital cost of $330 million in 1996 dollars (Bloomster et al., 1990, p. 12). With operating costs, the discounted storage cost (at a 4% discount rate and with a life of 30 years) would be about $1.15 per gram per year when the facility was full. This makes the cost of storage of plutonium generated from nuclear power production low compared to the cost of electricity. At $1.50/gram, storing plutonium would add only one-tenth of one % to energy costs (Berkhout et al., 1993, p. 171).

Further, this storage would be as secure as MOX - and it would entail much less transportation.

Surface storage of plutonium also implies continuous surveillance and maintenance (S&M) costs for the following items:

1. Maintaining plutonium handling facilities.
2. Transportation services.
3. Inventory and accounting charges.
4. Security and national regulatory costs.

Given the National Academy of Sciences estimate that plutonium can be stored for $2-$4 per gram per year, the S&M costs for 38.2 metric tons of plutonium would be between $76.4 million and $152.8 million per year in perpetuity. At a 10% discount rate, the present value of storage is equal to 10 times $76.4-$152.8 million, or between $760 million to $1.52 billion. These costs would change if the stock of plutonium declines due to burning or burial. For example, if this stock of plutonium declined by 3.3 metric tons/year after ten years, the present value cost of storage would be reduced to $300 million to $600 million (Rothwell, 1996, pp. 6-7).

The most recent example of a surface storage project, the fissile material storage unit being built in Russia at MAYAK with US assistance, will cost roughly $300 million. Adding an industrial-scale plant to convert plutonium pits to oxide might add another $100 million. Thus, the up-front cost of this storage facility is probably about $400 million (Bunn, 1996, pp. 4-5). Russian construction costs are notoriously low because of the condition of the Russian economy. A $400 million expense equals about one-thousandth of the Russian GDP. A comparable US expenditure would be about $5 billion.

Using surface storage requires obtaining a site, and this will likely result in considerable political opposition. Further, one form of monitored storage, the Monitored Retrievable Storage (MRS) program, is linked by statute to progress on a geologic repository. For this reason, it is unlikely that a MRS can be brought into operation in the near term to support DOE's commitment to begin accepting spent fuel in 1998 (Nuclear Wastes: Technologies for Separations and Transmutation, 1996, p. 16).

Mined geologic disposal

The Nuclear Waste Policy Act of 1982 (NWPA), as amended in 1987, commits the US to geologic isolation as the best long-term solution to the final disposition of waste. The NWPA requires that the costs of geologic disposition of materials be paid through a nuclear waste fee of one mil ($0.001) per kWh of electricity generated by nuclear power. In 1994, this fund reached about $8 billion (Nuclear Wastes: Technologies for Separations and Transmutation, 1996, pp. 1, 11, 15). As currently envisioned, the cost of the geologic disposition option should be about the

124

same as that for vitrification plus the cost for burial in tunnels - i.e., the cost of both operations. However, as the following paragraphs demonstrate, costs have increased greatly for the larger programs. Specific vitrification costs are provided in the vitrification section at (4) below.

Deep geological disposal is attractive because plutonium does not mix with water (Fetter, 1992, pp. 144-148). The two parameters of interest in securing repositories are the maximum anticipated tunneling advance rate and the maximum distance around a repository at which surface facilities for tunneling might be located. Based on current state of the art, a tunneling advance rate of 100 m/day is a credible upper bound in both granite and tuff using a tunneling machine. By contrast, initial operations at the Yucca Mountain Repository progressed quite slowly - the first 800 meters of tunnel took 6 months. After that, the tunnel has advanced an average of 30 meters per day (Peterson, 1995, pp. 433-434).

Tunnel lengths of 10 km are routinely constructed from a single surface point and drill and blast methods can achieve between 2.5 and 5 meters progress per day. Two recent projects, each about 9 km long, cost between $24 million and $52 million in 1996 dollars, or between $2.7 and $5.8 million per kilometer (Peterson, 1995, pp. 433-434). This cost estimate agrees well with the estimate, noted in a previous section, that a 2.5 meter hole 300 meters deep can be constructed for about $1.1 million in 1996 dollars (Management and Disposition of Excess Weapons Plutonium, 1994, p. 285).

Cost estimates for geologic repository disposal of spent fuel from commercial power reactors without vitrification were estimated to be about $400,000 per ton of heavy metal in 1996 dollars. However, the cost of disposal of plutonium would be higher because plutonium must be diluted to make re-extraction difficult. If this dilution option did not include vitrification of the resulting material, denaturing would cost on the order of several million dollars per metric ton of plutonium and total disposal costs would range from $130-$400 million for 50 metric tons of plutonium (Makhijani and Makhijani, 1995, p. 66).

In practice, the construction and use of long-term repository has been much more expensive than these estimates would imply. In addition, repositories have been far more politically contentious than originally estimated. From its start through April, 1995, the US government invested about $1.6 billion in the Waste Isolation Pilot Project (WIPP) for low level waste. It is still unopened. At the proposed Yucca Mountain repository for high level waste, costs have increased and delays have occurred partially because of a US Nuclear Regulatory Commission (NRC) regulation that requires the encapsulation of spent fuel to remain effective for at least 300 to 1000 years with longer limits, typically 10,000 years - on release rates (Nuclear Wastes: Technologies for Separations and Transmutation, 1996, pp. 1, 119). Given the hazardous nature of the materials involved, this regulation is quite reasonable and the increased costs that have resulted are measures of the cost of protecting the environment when geological disposal is used.

Even larger cost increases at Yucca Mountain have arisen from to poor engineering, poor project management by the DOE, and faulty science underlying the location and design of the project. As of 1996, the Yucca Mountain Repository was years behind schedule. Its costs were estimated at $22 billion with an additional $1 billion for a 10% capacity increase. Included in the $22 billion was $13 billion in development costs. The Nuclear Waste Policy Act limits loading of Yucca Mountain to 70,000 tons of heavy metal until a second repository has begun. However, an operating repository for civilian nuclear waste is still at least 14 years away and it is estimated that about $10 billion will be expended on that project even before a site is selected (Nuclear Wastes: Technologies for Separations and Transmutation, 1996, pp. 11, 12, 93).

Deep borehole disposal

This option was favorably reviewed by the National Academy of Sciences in their 1994 report. Material would be placed in the lower 2000 meters of holes that extend 4000 meters below the surface. Processing before emplacement would be required to insure proper use of space and to avoid nuclear chain reactions. Even if plutonium was only 10% of the final material placed in the borehole it would still be possible to put 50 metric tons of plutonium in a single hole (Management and Disposition of Excess Weapons Plutonium, 1994, pp. 257-263).

Boreholes with diameters of approximately 80-140 cm would be drilled in solid, crystalline rock. The canisters would have a diameter of 50 centimeters or larger and would be sealed from one another with bentonite clay. The top of the hole would be sealed with 2000 meters of bentonite clay, asphalt and concrete. Swedish estimates are that each hole would cost about $110 million in 1996 dollars. Russian estimates are much lower. It is assumed that in the United States the costs of development of this option would substantially surpass the actual costs of emplacement (Management and Disposition of Excess Weapons Plutonium, 1994, pp. 257-263).

Vitrification

Vitrification, mixing plutonium with melted glass, is one of the most heavily researched and one of the most technologically and economically certain disposition methods. As a result, vitrification was one of two options preferred by the NAS and chosen by the DOE in December, 1996 for plutonium disposition.

The French have worked with vitrification since 1957. In 1968, the PIVER vitrification unit began using batch or pot vitrification techniques. PIVER made about 200 glass blocks, each weighing 100 kg. Three current French plants have produced over 5000 canisters containing about 2000 tons of glass. The French view borosilicate glass to be safer although it is more difficult to use (Kushnikov, 1995, p. 27). They claim they could vitrify 8.5 metric tons of plutonium per year by directly feeding plutonium dioxide with fission products into the vitrification stage of the production process. This would make about 300 kg of glass for each 8 kg of plutonium (Jouan, 1995, pp. 360-363).

The first pilot Russian vitrification plant was built at MAYAK in 1987. The technology in this plant is based on electric heating of phosphate glass. The plant processed 9100 cubic meters of high level liquid waste, producing more than 1800 tons of phosphate class. A second vitrification production line was under construction in 1996 (Kushnikov, 1995, p. 27). Russian experiments have used non-borosilicate glass which has a life span of about 300 years (Diakov, 1996).

By 1994, the DOE had spent over $1 billion trying to vitrify liquid wastes and had not yet succeeded. Plutonium does not pose the problems that liquids do, but vitrification of plutonium alone would not present a sufficient barrier to reuse (Makhijani and Makhijani, 1995, p. 4). If plutonium disposition must be accomplished quickly, vitrification in the US may pose a problem. The Defense Waste Processing Facility at Savannah River was 6 years behind schedule when it opened in 1996, and a Westinghouse Savannah River scientist has stated that it could not vitrify plutonium for 15 years after being given the order to do so (Carter, 1994, p. 43). However, once a plant is operational, Makhijani has estimated that if plutonium was mixed at the 1% level, 50 metric tons of material could be vitrified in 7.5 to 15 years. If, instead, plutonium was mixed at the 5% level, 50 tons could be vitrified in as little as 1.5 to three years (Makhijani and Makhijani, 1995, p. 57).

Critics claim that vitrification leaves weapon-grade plutonium in a recoverable form that is not isotopically contaminated. They also claim that although plutonium is relatively insoluble in water, boron, which is the medium that absorbs neutrons, is not. Thus, they claim vitrified waste stored in the presence of water could go critical (DeVolpi, 1995, pp. 20-21). Russia has always perceived existing vitrification technology as being unsafe due to plutonium settling problems that could lead to criticality (Kushnikov, 1995, p. 27).

Research on "clean glass" borosilicate vitrification performed by an MIT engineering group found that extracting plutonium from this medium would present a formidable barrier to subnational groups. Tests conducted at MIT also showed borosilicate glass to be quite durable in the presence of water (Wenzel et al., 1996, pp. 59-60). Further, the fact that plutonium can be re-extracted by the major nuclear powers may be an advantage when it comes to getting countries that view plutonium as a long-run energy source to sign international agreements on vitrification.

There are three general options to make plutonium extraction more difficult for subnational groups. Two of these options involve either mixing plutonium with elements like gadolinium or thorium that are difficult to separate chemically or mixing plutonium with fission products from spent fuel to poison the mixture (Makhijani, 1995, pp. 276-277). These two general options have led to three specific approaches to mixing plutonium:

1. Vitrification of plutonium mixed with gamma-emitting fission products so the resulting glass logs meet the spent fuel standard. These fission products have much shorter half-lives than plutonium. For example, the half-life of Cesium 137 is only 30 years as opposed to 24,000 years for plutonium. Thus, this

mix would become less resistant to proliferation over time. Using this mix is likely to delay vitrification since plants are not prepared for this task (Makhijani and Makhijani, 1995, p. 88). This option makes it easier to extract the plutonium at a later date and once the plants have been built, it would allow the vitrification of 50 tons of plutonium in 400 tons of glass in two years (Jouan, 1995, pp. 360-363).

2. Vitrification of plutonium with depleted uranium or some other alpha-producing element.

3. Vitrification of plutonium with a non-radioactive element, such as europium, that would render the mixture unsuitable for weapons without reprocessing (Makhijani and Makhijani, 1995, p. 4).

A third option would create a barrier to misuse by subnational groups by making the canisters in which vitrified plutonium is stored highly radioactive (Makhijani and Makhijani, 1995, p. 89). Such a steel canister might be poisoned with Cesium 137. None of the radiation from the Cesium would be absorbed by the glass (Makhijani, 1995, pp. 276-277).

These options were expanded in a 1995 Programmatic Environmental Impact Statement/Record of Decision (PEIS/ROD) on plutonium disposition that suggested the following five methods of glass or ceramic immobilization.

1. Vitrification with an internal radiation barrier - two options:
   a. Use a new facility that produces borosilicate glass containing plutonium, neutron absorbers, and Cesium 137 as a radiation barrier, and then encapsulates this glass in a storage container.
   b. Use an adjunct melter with an existing Defense Waste Processing Facility (DWPF) that produces a glass containing plutonium, neutron absorbers, and high level waste, and then encapsulates this glass in a storage container.

2. Vitrification with an external radiation barrier.
   Employs a can-in-canister variant, in which an inner can containing a plutonium- and neutron-absorber-bearing glass is surrounded by a glass containing a radiological barrier, which, in turn, is contained in an outer storage canister.

3. Ceramics with an internal radiation barrier.
   Uses a new facility that produces a ceramic containing plutonium, neutron absorbers, and Cesium 137, and then encapsulates the ceramic in a storage canister.

4. Ceramics with an external radiation barrier.
   Employs a can-in-canister alternative in which an inner can of a ceramic containing plutonium and neutron absorbers is surrounded by a ceramic or glass that contains a radiological barrier, which is in turn contained in an outer storage canister (Gray et al., 1995, pp. 60-61).

The planning assumptions for these kinds of immobilization and the construction of new plants they would require assume it would take ten years to process 50 metric tons of plutonium (Kan and Sullivan, 1995, p. 296). According to a June, 1996, report from the DOE, "as a packaging strategy, co-disposing any of the waste forms with packaged vitrified HLW appears technically sound and cost-effective." (Technical Strategy for the Treatment, Packaging, and Disposal of Aluminum-Based Spent Nuclear Fuel, 1996, p. 4) In addition, vitrification is unlikely to experience the problems that gallium contamination of surplus weapon plutonium poses for MOX fabrication (Toevs and Beard, 1996). Since the prime objective of disposition is to remove from circulation those materials that are directly usable in weapons, this attribute would seem to strongly favor a vitrification option.

Fabrication costs of vitrification Several cost estimates are available for vitrification plants. In some cases these plants would handle plutonium only. In others, the estimates include wider options that would deal with general nuclear waste disposal. As the following estimates show, the high capital cost of vitrification plants coupled with the low marginal costs of production mean that those plants that can increase the total output of vitrified material will enjoy a lower cost per canister based on significant economies of scale.

According to a 1990 proposal, the US could incorporate plutonium into 25,000 tons of glass at a rate of about 1000 tons of glass per year. This would allow the disposal of 100 tons of plutonium in five years if the glass contained only 2% plutonium. An analysis by Pacific Northwest Laboratories estimates the total additional cost to convert 100 tons of plutonium metal to oxide and vitrify it with other high level waste at $120 million ($1996) (Fetter, 1992, pp. 144-148). This estimate equates to about $1.20 per gram of plutonium, and compares favorably with surface storage options. However, no costs of storage or separation are included in this estimate and the costs of vitrification are extremely low compared to other recent estimates.

A 1993 study of vitrification costs at Hanford estimated that the capital cost of a waste vitrification plant would be about $1.85 billion in 1996 dollars. The plant could produce 372 canisters of vitrified material per year for 40 years, yielding 14,880 canisters. Labor figures were not calculated, but the vitrification plant at the Savannah River Site has 1,150 on its staff. Assuming a staff of 1100 with an average wage of $45,000, the total labor cost over the life of the plant would be about $2 billion (Nuclear Wastes: Technologies for Separations and Transmutation, 1996, pp. 191-192). Thus, the cost of labor and capital alone would be about $260,000 per canister. When the remaining costs of canister production are added, this estimate corresponds well with the fabrication cost estimate of $300,000 per canister by the NRC discussed later in this section.

A 1996 DOE report estimated that using vitrification to dispose of all US surplus weapons plutonium would cost approximately $1.8 billion (Technical Summary Report For Surplus Weapons-Usable Plutonium Disposition, 1996, pp. 4-10). Given the relative amounts of weapon-grade material compared to the total volume of plutonium scrap, this estimate corresponds closely to a 1995 estimate by a team of LLNL researchers that an expenditure of $4.9 Billion to $5.8 Billion would be necessary to vitrify and dispose of all scrap plutonium whose purity ranged from 100% to .001% (Gray et al., 1995, p. 63).

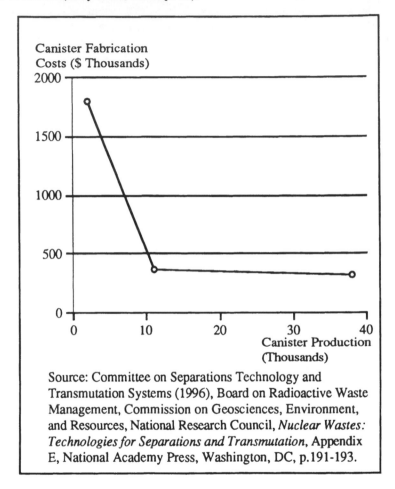

Source: Committee on Separations Technology and Transmutation Systems (1996), Board on Radioactive Waste Management, Commission on Geosciences, Environment, and Resources, National Research Council, *Nuclear Wastes: Technologies for Separations and Transmutation*, Appendix E, National Academy Press, Washington, DC, p.191-193.

Figure 8.1 The costs of canister fabrication at various production rates

A 1996 study of nuclear waste vitrification costs by the National Research Council displayed in Figure 8.1 shows how important economies of scale are in the determination of canister production costs. This study calculated that vitrification costs in 1996 dollars would be about $300,000 per canister at a production rate of 38,000 canisters over 30 years, compared to a cost of $1.8 million per canister if only 2000 canisters were made (Nuclear Wastes: Technologies for Separations and

130

Transmutation, 1996, p. 93). Figure 8.1 demonstrates that canister fabrication costs fall dramatically as the production of a vitrification plant increases up to 10,000 canisters. Beyond that point, additional reductions in canister costs are relatively insignificant. This graph provides the rationale for concentrating all vitrification activities at one site instead of building vitrification plants at each site where plutonium must be disposed of.

Storage costs for vitrified material   If one assumes that the cost of a passively air-cooled high level waste storage area is comparable to the dry storage required for vitrified material, total capital and variable costs in 1996 dollars for a vitrified high level waste storage area would range from $89,000-$177,000 /ton of heavy metal (Berkhout, 1993, p. 14). The 1996 NRC study estimated that these storage costs would be between $185,000 and $200,000 per canister, irrespective of the material that was vitrified (Nuclear Wastes: Technologies for Separations and Transmutation, 1996, pp. 191-193). Assuming that the previous fabrication cost estimates of about $300,000 per canister are correct, the total cost of vitrifying and storing plutonium would be about $500,000 per canister.

Summary - a comparison of disposition costs

Table 8.2 includes a rough comparison of disposition costs for the options covered in this chapter. Each of these costs has numerous caveats, and the reader is encouraged to carefully study the expanded sections of this chapter, as well as Chapters Five, Six and Seven, before drawing any conclusions about these comparisons. These costs generally apply for the specific option only and do not include the costs of ancillary reprocessing, the costs of environmental damage that may be caused by the option, or the development and licensing costs that would have to be incurred if the option was adopted.

131

Table 8.2
Summary comparison - cost estimates of disposition options billions of 1996 dollars

Non-viable Options	Costs	Amount of Pu
Fissioning By Nuclear Explosion	2.2	all surplus weapon
Burning In LWRs	1.8-2.4* 7.7-11.7**	not clearly specified
Burning In Fast Reactors	5.6-6.0	50 MT
Burning In Unconven. Matrices	1.8-2.4	50MT?
Launching Into the Sun	29	250MT
Sub-Seabed Disposal	Several billions	all surplus weapon
Dilution	Tens of millions[+]	all surplus weapon
Tectonic Plate Burial	Several billions	all surplus weapon
Mix and Melt Disposition	4	50MT
Transmutation	140 (accelerator transmutation)	90% of LWR actinides
Viable Options	**Costs**	**Amount of Pu**
Direct Surface Storage Disposal	3.2 7.5	200MT@$1.60/g/yr storage cost 200 tons@$4.00/g/yr storage cost
Deep Geologic Disposal	+2.7-5.8 million/kilometer of tunnel plus 1.8 .13-.4 or 22	250MT vitrified 50 MT not vitrified 70,000MT HLW at Yucca Mountain
Deep Borehole Disposal	.11 +.2-.3 for development	50 tons of Pu
Vitrification	1.8	250MT

*Existing LWRs. Add $200 million for gallium removal.
**Evolutionary LWRs. Add $200 million for gallium removal.
[+]This option probably has the highest cost to the environment of any option

Note

1   The spent fuel standard proposes to make plutonium as difficult to retrieve as it would be if it was in the form in which it exists in nuclear reactor fuel that has been irradiated (used) to the extent that it can no longer effectively sustain a chain reaction and thus, has been removed from the reactor for disposal. This irradiated fuel contains fission products, uranium, and transuranic isotopes.

References

Berkhout, F., Diakov, A., Feiveson, H., Hunt, H., Lyman, E., Miller, M., von Hipple, F. (1993), 'Disposition of Separated Plutonium', *Science and Global Security*, Volume 3, pp. 169-171

Berkhout, Franz (1993), *Fuel Reprocessing At THORP: Profitability and Public Liabilities*, Center for Energy and Environmental Studies, Princeton University, Princeton, NJ, pp. 8, 14.

Bloomster, C.H., Hendrickson, P.L., Killinger, M.H., Jonas, B.J. (1990), *Options and regulatory issues related to disposition of fissile materials from arms reduction*, PNL-SA-18728, Pacific Northwest Laboratories, US Department of Energy, Washington, DC, pp. 12-13.

Bunn, Matthew (1996), *Getting the Plutonium disposition Job Done: The Concept of a Joint Venture Disposition Enterprise Financed by Additional Sales of Highly Enriched Uranium*, International Conference on Military Conversion and Science: Utilization/Disposal of the Excess Weapon Plutonium: Scientific, Technological and Socio-Economic Aspects, Como, Italy, March 18-20, 1996, pp. 4-5.

Campbell, Ronald L. and Snider, J. David (1996), *Cost Comparison for Highly Enriched Uranium Disposition Alternatives*, Y/ES-122, Nuclear Materials Disposition Program Office, Y-12 Defense Programs, Lockheed Martin Energy Systems, Inc., p. 5.

Carter, Luther J. (1994), 'Let's Use It', *The Bulletin of the Atomic Scientists*, Vol. 50, No. 3, p. 43.

Developing Technology to Reduce Radioactive Waste May Take Decades and Be Costly (1993), GAO/RCED-94-16, United States General Accounting Office, Washington, DC, pp. 3-13.

DeVolpi, Alex (1995), 'Fast finish to plutonium peril', *The Bulletin of the Atomic Scientists*, Vol. 51, No. 5, pp. 20-21.

Diakov, Anatoli S. (1996), *Utilization of Already Separated Plutonium in Russia: Consideration of Short- and Long-Term Options*, International Conference on Military Conversion and Science: Utilization/Disposal of the Excess Weapon Plutonium: Scientific, Technological and Socio-Economic Aspects, Como, Italy.

Fetter, Steve (1992), 'Control and Disposition of Nuclear Weapons Materials', *Working Papers of the International Symposium on Conversion of Nuclear Warheads for Peaceful Purposes*, Rome, Italy, pp. 144-148.

Garwin, Richard L. (1993), *The Use of Pu for Peaceful Purposes*, Presentation at Cornell University.

Gray, L., Kan, T., Shaw, H., Armantrout, G. (1995), 'Immobilization Needs and Technology Programs', *Final Proceedings: US Department of Energy Plutonium Stabilization and Immobilization Workshop*, pp. 60-63.

Hardung, Heimo (1994), 'Plutonium Naiveté', *Defense News*, April 4-10, pp. 22, 24.

Johnson, E.R. (1992), *Alternatives for Disposal of Plutonium from Nuclear Weapons Disarmament Activities*, Annual Meeting of the Institute of Nuclear Materials Management, Orlando, Florida.

Jouan, Antoine (1995), 'Selecting a Plutonium Vitrification Process', *Final Proceedings: US Department of Energy Plutonium Stabilization and Immobilization Workshop*, pp. 360-363.

Kan, Tehmau and Sullivan, Kent (1995), 'Glass and Ceramic Immobilization Alternatives and the Use of New Facilities', *Final Proceedings: US Department of Energy Plutonium Stabilization and Immobilization Workshop*, p. 296.

'Known Nuclear Tests Worldwide', *1945-1994* (1995), *The Bulletin of the Atomic Scientists*, Vol. 51, No. 3, p. 70.

Kushnikov, Viktor (1995), 'Radioactive Waste Management and Plutonium Recovery Within the Context of the Development of Nuclear Energy in Russia', *Final Proceedings: US Department of Energy Plutonium Stabilization and Immobilization Workshop*, p. 27.

Leslie, Bret (1994), 'DOE Assesses Dangers from its Plutonium Inventory', *Science for Democratic Action*, Vol. 3, No. 2, IEER, Washington, DC, pp. 2-3.

Lombardi, Carlo and Mazzola, A. (1994), *Plutonium Burning Via Thermal Fission in Unconventional Matrices*, IAEA Technical Committee Meeting on Unconventional Options for Plutonium Disposition, Obninsk, Russia.

Lombardi, Carlo (1994), Dipartimento Di Ingegneria Nucleare, Centro Studi Nucleari Enrico Fermi, *Letter to William Weida*, November 30.

Lombardi, C., Mazzola, A., Vettraino, F. (1996), *Non-Fertile Fuels for Burning Weapons Plutonium in Thermal Fission Reactors*, International Conference on Military Conversion and Science: Utilization/Disposal of the Excess Weapon Plutonium: Scientific, Technological and Socio-Economic Aspects, Como, Italy, pp. 5-6.

Makhijani, Arjun and Makhijani, Annie (1995), *Fissile Materials In A Glass, Darkly*, IEER Press, Takoma Park, Maryland, pp. 4-100.

Makhijani, Arjun (1995), 'An Approach to Meeting the Spent Fuel Standard', *Final Proceedings: US Department of Energy Plutonium Stabilization and Immobilization Workshop*, pp. 276-277.

Management and Disposition of Excess Weapons Plutonium (1994), Committee on International Security and Arms Control, National Academy of Sciences, National Academy Press, Washington, DC, 1994, pp. 10-285.

Nuclear Wastes: Technologies for Separations and Transmutation (1996), Committee on Separations Technology and Transmutation Systems, Board on Radioactive Waste Management, Commission on Geosciences, Environment, and Resources, National Research Council, National Academy Press, Washington, DC, pp. 1-193.

Panofsky, Wolfgang K.H. (1996), 'No Quick Fix For Plutonium Threat', *The Bulletin of the Atomic Scientists*, Vol. 52, No. 1, pp. 3, 59.

Peterson, Per F. (1995), 'Long-Term Retrievability and Safeguards for Immobilized Weapons Plutonium in Geological Storage', *Final Proceedings: US Department of Energy Plutonium Stabilization and Immobilization Workshop*, pp. 433-434.

Plutonium Fuel: An Assessment (1989), OECD/NEA, Paris, Table 14.

Plutonium - Deadly Gold of the Nuclear Age (1992), International Physicians for the Prevention of Nuclear War and The Institute for Energy and Environmental Research, International Physicians Press, Cambridge, MA, pp. 130-138.

Protection and Management of Plutonium (1995), American Nuclear Society Special Report, p. 10.

Rothwell, Geoffrey (1996), *Economic Assumptions for Evaluating Reactor-Related Options for Managing Plutonium*, International Conference on Military Conversion and Science: Utilization/Disposal of the Excess Weapon Plutonium: Scientific, Technological and Socio-Economic Aspects, Como, Italy, pp. 6-7.

Schulze, Joachim (1992), 'Burning of Plutonium in Light Water Reactors (MOX Fuel Elements) Compared To Other Treatment', *Working Papers of the International Symposium on Conversion of Nuclear Warheads for Peaceful Purposes*, Rome, Italy, pp. 65-74.

Schwartz, Steven I. (1997), Ed., *Atomic Audit*, Brookings, Washington, DC, pending publication.

Taylor, Theodore B. (1996), *Solar Disposal of Radioactive Wastes and Plutonium*, Submitted to Science and Global Security, pp. 24-25.

Technical Strategy for the Treatment, Packaging, and Disposal of Aluminum-Based Spent Nuclear Fuel (1996), Office of Spent Fuel Management, Department of Energy, pp. 4-10.

Tigner, Brooks (1994), 'Nuclear Bomb Experts: Detonate Stockpile Underground', *Defense News*, September 4, p. 10.

Toevs, J. and Beard, C. (1996), *Gallium in Weapons-Grade Plutonium and MOX Fuel Fabrication*, LA-UR-96-4764, Los Alamos National Laboratory, NM.

Uematsu, Kunihiko (1992), 'The Technological and Economic Aspects of Plutonium Utilization in Fission Reactors', *Working Papers of the International Symposium on Conversion of Nuclear Warheads for Peaceful Purposes*, Rome, Italy, pp. 109-115.

von Hipple, Frank (1992), 'Control and Disposition of Nuclear Weapons Materials', *Working Papers of the International Symposium on Conversion of Nuclear Warheads for Peaceful Purposes*, Rome, Italy, pp. 119-128.

Wenzel, K.W., Sylvester, K.W., Cerefice, G. (1996), "Reply to 'Fast finish to plutonium peril'", *The Bulletin of the Atomic Scientists*, Vol. 52, No. 1, pp. 59-60.

9 Conclusion: the major cost drivers in disposition

Introduction

Some studies on alternative disposition methods for both plutonium and HEU claim the danger proliferation problems pose for all people obligates decision makers to select among disposition options based on expediency and safety, and that economic considerations should not play a major role in this process (For example, see Makhijani and Makhijani, 1995). However, a student of either the military budgeting process or the budget considerations surrounding a major infectious disease such as AIDS will realize that there is no precedent for real-world decisions - even those that concern threats to large numbers of people - being made in an environment free of economic considerations. In fact, in making such decisions it is not unusual for economic costs and benefits to be considered first, not last. For this reason, it is necessary to identify those factors in the disposition area that will create common costs across all options, and to specify those areas where specific factors are likely to be major cost drivers that could discriminate between the various disposition options.

HEU disposition

This book has shown that while highly enriched uranium (HEU) can be down-blended and burned by nuclear reactors for power generation, it will face the same long-term economic forces as the nuclear industry in general. As a result, all other issues aside, downblending is unlikely to be financially successful in the United States if it is planned as a long-term program. Unless downblended HEU can be moved to the front of the pipeline and disposed of while there are sufficient reactors remaining in the United States, the US may be left with substantial amounts of HEU for which there is no demand. As of 1996, the operations of US Enrichment Corporation, the entity that handles downblended fuel, have not been helpful - for purely economic reasons - in significantly reducing the outstanding stocks of HEU.

Current HEU disposition programs designed by the Department of Energy are predicated on a positive financial return to the US government. Since this seems to be unrealistic in the long run, other goals may have to be established. For example, the US may have to apply the same standards to HEU disposition as it applies to plutonium. Insistence on using the economic demand for reactor fuel as the sole driver of the HEU program is likely to create a large amount of weapon-grade or downblended HEU for which there is no economically viable reuse program and for which there are no other planned disposition options.

Plutonium disposition

This paper has shown that burning plutonium in commercial reactors is not economical viable and would require significant subsidies. As the 1994 National Academy of Sciences study stated:

> Exploiting the energy value of plutonium should not be a central criterion for decision making, both because the cost of fabricating and safeguarding plutonium fuels makes them currently noncompetitive with cheap and widely available low-enriched uranium fuels, and because whatever economic value this plutonium might represent now or in the future is small by comparison to the security stakes (Management and Disposition of Excess Weapons Plutonium, 1994., pp. 3-4).

However, even if burning plutonium is not economical, a more important question is whether it is still cheaper than other methods of dealing with or disposing of plutonium.

This question incorporates both proliferation risk and economics, and it is unlikely to be answered by the current DOE methods of cost analysis. For example, according to a 1995 report on melter technologies, the relevant costs to be considered before deployment of the technology should be based on:

1. Lifecycle costs which include equipment/initial costs, operations, and maintenance costs.
2. Estimated development/adoption costs.
3. Consumable equipment cost, measurable by the number of melters during lifetime.
4. Decommissioning cost (Perez et al., 1995, p. 345).

Important as these costs may be, if they are treated in this simple accounting sense they cannot answer the significant tradeoff questions posed in the preceding paragraph. The following framework is suggested as an alternative way in this question might be considered:

1.	Increased handling of plutonium leads to increased costs and increased proliferation risks.
2.	Any proposal to burn plutonium in reactors to reach a spent fuel standard might also be accomplished more simply and cheaply by mixing plutonium with waste to a spent fuel standard to start with (Panofsky, Wenzel et al, and DeVolpi, 1996). As an isotopically different element, plutonium can always be chemically separated from spent fuel whether it was generated inside a reactor or simply mixed with existing spent fuel. The difficulty associated with this operation can be increased by adding other elements to the mix.
3.	Waste storage costs, irrespective of the storage method chosen, are based on volume - assuming radioactivity is roughly identical across all burning and non-burning options. In any process that requires putting material in a reactor, whether for power generation or simply to dispose of the material, the total volume of material will remain constant throughout the process. The only exception to this rule occurs when reprocessing is involved. Then both waste volume and costs rise dramatically.
4.	Costs are altered for transmutation because although one is handling hotter material for relatively shorter periods of time, these time periods are still so extensive that discounted cost comparisons between alternatives cannot show significant differences. In addition, transmutation technologies still require reprocessing and they still must absorb the cost of research and development. Other options do not have either of these negatives.

Viewed in this light, final waste disposal costs will be incurred whatever disposal option is taken. These costs could potentially be offset by doing something profitable with plutonium and HEU prior to final storage, but this book has shown that the demand for downblended HEU is directly dependent on the health of the nuclear industry while finding a profitable use for plutonium is highly unlikely. Thus, with the possible exception of some HEU, the more probable case is one where the costs of disposition are not offset by other revenues and are, instead, increased by the unique costs associated with the disposition option chosen. The factors most likely to significantly increase costs are the major cost drivers that create differences among the various options for plutonium and HEU disposition At this point, major costs appear to arise from four areas:

1.	The level of subsidization in the 'profitable' parts of the disposition program.
2.	Those items (such as reprocessing) that increase the volume of waste and thus, the cost of waste disposal.
3.	The cost of security and its direct relationship to the number of times a material is handled or moved.
4.	The cost of research and development of new and unproved methods of disposition.

These four costs appear to outweigh all other costs of disposition by many orders of magnitude. Minimizing their impact should be a major consideration in choosing among disposition options, and there is historical evidence to support of this approach. A 1981 RAND study of the factors leading to capital cost overruns and performance shortfalls in pioneer chemical projects similar to those required for disposition found the following significant problem areas:

1. Severe underestimation of capital requirements is the norm for all new technologies.
2. Capital costs are repeatedly underestimated for advanced chemical process facilities just as they are for advanced energy process plants. Further, performance of advanced plants falls far short of expectations.
3. Most variation in cost estimation error can be directly explained by:
   a. The extent to which the methodology departs from previous plants.
   b. The degree of definition of the project's site and related characteristics.
   c. The complexity of the plant.
4. Most of the variation in plant performance is explained by the amount of new technology and whether a plant produces solid materials (Merrow et al., 1981).

In addition to these issues, many non-technical factors have unnecessarily increased the costs of disposition. These factors have the potential to cause significant future disposition cost increases unless they are carefully controlled. Many of these factors are directly related to the manner in which DOE operates and the following 'common sense' suggestions are proposed as a way of keeping these factors in control:

1. DOE must not compromise proliferation goals for money. 'Proliferation risk' should be combined with 'commercial use' in evaluating options.
2. Absent a credible treatment and disposal plan for spent fuel, DOE should not create more spent fuel.
3. The 'spent fuel standard' establishes a minimum level of proliferation resistance. It should not be used to encourage the creation of spent fuel, nor should it be interpreted as establishing a goal for the disposition of weapons usable radioactive materials.
4. Disposition must not be viewed as a short-term solution to a long-term problem.
5. Disposition decisions must not compromise the health and safety of workers, the public, or the environment.
6. DOE must adopt a 'cradle to grave' analysis of environmental impacts, as the National Environmental Policy Act requires. If spent fuel is generated as a result of disposition, that spent fuel must be accounted for from its creation to its disposition.
7. Upon being declared surplus, weapons usable radioactive materials should be placed under international (IAEA) control.

8. All disposition activities should conform to international standards of safeguard and transparency.
9. Ideally, disposition decisions should have results that are irreversible.
10. All disposition decisions should be reflexive - they should mirror the actions we demand, expect, or desire of other nations.

References

Makhijani, Arjun and Makhijani, Annie (1995), *Fissile Materials In A Glass, Darkly*, IEER Press, Takoma Park, Maryland.

Management and Disposition of Excess Weapons Plutonium (Pre Publication Copy) (1994), Committee on International Security and Arms Control, National Academy of Sciences, National Academy Press, Washington, DC, pp. 3-4.

Merrow, E.W., Phillips, K.E., Meyers, C.W. (1981), *Understanding Cost Growth and Performance Shortfalls in Pioneer Process Plants*, RAND/R-2569-DOE, RAND Corporation, Santa Monica, California.

Panofsky, W., Wenzel, K. et al, DeVolpi, A. (1996), 'Letters', *The Bulletin of the Atomic Scientists*, vol. 52, no. 1, January/February.

Perez, J.M., Schumacher, R.F., Forsberg, C.W. (1995), 'Melter Technologies Assessment', *Final Proceedings: US Department of Energy Plutonium Stabilization and Immobilization Workshop*, p. 345.

Appendix 1

Cases where economic conversion serves as a rationale for selecting specific methods of plutonium disposition

Introduction

When defense nuclear facilities closed at the end of the Cold War, the isolated regions that depended on these facilities for their economic livelihood often found they were left with large amounts of contamination and little or no possibility of attracting replacement industries. Several of these regions were later approached by the DOE, its contractors, or by British or French companies (BNFL or COGEMA) with the following proposition: allow the old, defense nuclear site to be converted to DOE's 'new mission' of plutonium disposition through MOX fabrication, MOX burning, or any of a number of direct disposal options. These offers were usually attractive to citizens of a region where government cutbacks were occurring because:

1. The region is accustomed to living next to a DOE nuclear facility.
2. No one else is interested in the DOE site due to contamination.
3. The economic future of one or more communities may be at stake.

Conversion is a post-W.W.II concept based on the assumption that similarities between defense and civilian technologies and facilities will lead to reuse of former defense sites. This is not a valid assumption when nuclear weapons are involved and hence, there has never been much hope that production facilities for nuclear weapons would be converted to something else. Viewed purely from an economic standpoint, alternative uses for closed defense nuclear facility are unlikely to be found because defense nuclear resources - and many civilian nuclear resources - have no other economic uses. Therefore, the money spent on these resources is generally a sunk cost. For example, resources devoted to past nuclear weapon development and construction have been lost and cannot be recovered. Further, a large share of the annually appropriated resources previously used for deployment and operations of nuclear weapons, instead of becoming available for other uses at the end of the Cold War, are now consumed by:

1. Continuous, low level nuclear weapon operations - 60% of the national laboratory budgets are still devoted to nuclear weapons. In addition, US plans to maintain about 3500 nuclear warheads in the future and to have a number of larger defense nuclear facilities remain open may consume three quarters of the present nuclear budget indefinitely.
2. Rising cleanup costs from past nuclear operations - these costs were $6.5 billion in FY1995 and total cleanup costs are likely to range from $400B to $1 Trillion by the time the entire defense nuclear complex is cleaned up.
3. The cost of dismantling warheads and disposing of new waste.

As a result, any community that wants to assure continued DOE spending in its region must find a way to associate itself with one of these three expenditure categories. In this respect, reprocessing, fabricating MOX, or burning MOX may be justified by marketing these processes as a tool of economic conversion and trying to gloss over the fact that reprocessing or MOX burning is not economically viable.[1] A proposal of this nature raises valid issues concerning the purpose of economic conversion and, if conversion is desirable, whether a conversion program of similar impact could be provided more cheaply.

Conversion as a general principle

In a time of restricted federal spending, no regional conversion program is likely to receive federal support if that program entails a significant, long-term federal budget obligation. In other words, unless a potential conversion program can demonstrate a net economic return that attracts private investors, it is highly unlikely that a proposed conversion program will become a reality and act as a substitute for the abandoned federal project it replaces. After the hurdle of basic economic feasibility is cleared, it is also necessary to insure that decisions about the choice of facilities (either nuclear or other large facilities) to be used as a vehicle for economic conversion meet with the full concurrence of the communities in which the projects will be built and the waste deposited. And finally, a full accounting of all benefits and costs, including waste disposal and long-term waste or spent fuel storage, should be made before any conversion project is approved (Weida, 1994).

When these basic criteria have been satisfied, both the region and the nation must consider the effects of a proposed project on the fundamental purposes of economic conversion:

1. On a site basis, the purpose of conversion is to preserve the economic community around the site by changing the economic base (source of external funds) for the site.

2.   For an entire economic sector, the purpose of conversion is to free resources to
      be used in other ways to benefit the community and nation.

These purposes can only be achieved if conversion generates economic benefits, not
liabilities.   With this in mind, the purpose of conversion is not to substitute one
government funded program for another - it is to change the region's economic base.
This cannot be achieved unless conversion generates a net economic benefit.
   The Isaiah and Triple Play proposals demonstrate how advocates of certain
plutonium disposition options have tried to use site conversion to adapt to these
economic realities. The Triple Play Reactor, proposed for the Savannah River Site
(SRS), and Project Isaiah, proposed for the old Washington Public Power System
(WPPS) reactors around the Hanford site, have both been suggested as conversion
programs where new or refurbished reactors would burn plutonium.   Both programs
have claimed they would be privately financed and, by implication, profit-making.

The Isaiah project

This project was proposed in 1993 to burn plutonium in MOX and to produce
electricity by completing the WPPS #1 reactor at Hanford, Washington, and the #3
reactor at Satsop, Washington.   It has been claimed the Isaiah project would create
9,000 direct construction jobs, 2,500 permanent operations jobs and 13,500
secondary jobs in the Hanford and Satsop regions. Each plant would produce 1,300
Mwe (Wages, 1993).
   In 1996 dollars, completion costs for WNP-1 were estimated at $1.85 billion, and
for WNP-2 they were calculated to be $1.75 billion. Operating costs were estimated
at about $23 million/year, and O&M costs at about $134 million/year including the
spent fuel disposal fee. When financing costs were included, the completion cost for
WNP-1 rose to $3.0 billion. However, private financing was supposed to cover all
project completion costs and, in addition, return $4.35 billion to the Federal
government (Honenkamp, 1993).
   These financial arrangements, and the employment they produced, sounded
promising when the Isaiah project was initially proposed, and it was supported as an
economic conversion tool by both the Washington construction and trade unions and
the civic leadership in the Tri-Cities area of Richland, Kennewick and Pasco,
Washington. However, calculations of power demand and supply, as well as the
potential revenues from this type of power generation, led to such marginal
economic projections that the financial outlook for the project was summed up by a
clause in the Project Isaiah contract that stated that DOE would "enter into a long
term contract......[with] *a federal obligation to make debt service payments if
revenues from the sale of steam [power is] not adequate.*" (Dodd, 1993) [Author's
Italics] These unfavorable economic projections, and the likelihood that the federal
government would eventually become liable for Isaiah's debt service payments, both
violated the principle that a conversion project must produce sufficient net economic
benefits to avoid encumbering the federal government with long-term financial

obligations. This was a significant factor in the termination of interest in this project.

The triple play reactor

After the DOE stopped producing tritium for nuclear warheads in the late 1980's, employment cutbacks began to affect employment in the region around the Savannah River Site (SRS), near Aiken, SC. A quasi-private consortium expressed interest in using either the advanced light water reactor (ALWR) or the modular high temperature gas-cooled reactor (MHTGR_ in a 'triple play' mode where the reactor would burn plutonium, and produce both tritium and electricity at the same time. The DOE had spent some $650 million on the MHTGR and $425 million on ALWR technology by 1994 as part of its nuclear energy and new production reactor programs. After the MHTGR failed as an electrical generator at a civilian nuclear power plant in Ft. St. Vrain, Colorado, the ALWR became the reactor of choice for the triple play project.

While existing reactors could have been modified to burn MOX fuel and produce tritium, the consortium proposed that new reactors of System 80+ design be built to burn plutonium, produce tritium and generate electricity at SRS. The System 80+ 'triple play' reactor was an advanced, pressurized light water reactor that used a 100% MOX core and produced 1350 MWe. Two units would consume 100 metric tons of plutonium in 30 years and would cost $6.6 billion ($1996) to build and deploy (Program Plan for Deployment of a System 80+ Multi-purpose Nuclear Facility at Savannah River Site, 1994, pp. 1-12).

Combustion Engineering, a subsidiary of Asea Brown Boveri, claimed two new reactors of this design would consume the surplus American plutonium stockpile while generating electricity that could be sold to the utilities. However, George A. Davis, the company's project manager for South Carolina, claimed the reactors did not make sense as an energy project, noting that "nobody in their right mind would put up with the risks and headaches just to generate electricity." But Davis also claimed the reactors were still a relatively cheap way to get rid of plutonium. And although the resulting spent fuel would still have about 80% of the plutonium contained in the original fuel, after removal from a reactor this plutonium would meet the spent fuel standard (Wald and Gordon, 1994).

In spite of this, the System 80+ approach appeared to be contrary to national non-proliferation objectives because the US has a long-standing policy of separating civilian nuclear energy production from nuclear weapons manufacturing. A multi-purpose reactor would blur this distinction and might strengthen the case of other countries who decided to use their electricity-producing nuclear facilities, as India had done, to make nuclear weapon materials. Further, there was an inherent contradiction in using a new reactor to produce tritium for weapons and to dispose of plutonium for weapons.

The System 80+ reactor's Program Plan displayed considerable "uncertainty in costs" in MOX fabrication and the plan proposed that the US federal government provide $50 million in up-front financing (Program Plan for Deployment of a System 80+ Multi-purpose Nuclear Facility at Savannah River Site, 1994, pp. 8-9). The private consortium offered to pay back the money if DOE ultimately decided to proceed with the System 80+ proposal at the end of the three-year study phase (Davis, 1995).

In addition, the Triple Play reactor Program Plan required an extensive list of other subsidies:

1. The federal government had to provide a site and infrastructure at no cost to the consortium.
2. The consortium paid disposal fees for waste, but then passed them through to the government, not to the consumer of the power.
3. The government supplied plutonium oxide, depleted uranium oxide, and the site lease, all at no charge, and it further agreed to sole-source irradiation services from the plant.
4. The 'annual fees' required from the government were estimated at $78 million for plutonium burning alone - about a 10% subsidy.
5. An annual fee would also be assessed for tritium production based on revenue losses and other factors.
6. The government shared liability for any increased costs due to regulatory changes or any other factors over which the consortium had no control (Program Plan for Deployment of a System 80+ Multi-purpose Nuclear Facility at Savannah River Site, 1994, pp. 68-75 and Davis, 1995).

As the calculations in Chapter Six showed, similar subsidies would probably have been required by the Isaiah project because a majority of the proposed revenues from both projects are from electrical generation. An electricity-producing, plutonium-burning light water reactor is not economically feasible because of the additional facilities and security procedures required for plutonium handling. MOX fabrication also adds hundreds of millions of dollars to normal operating costs. Each of these factors increases the financial risk associated with the project.

In an October 10, 1995, DOE briefing on its 'Dual Track' strategy for producing tritium, Secretary O'Leary essentially abandoned the Triple Play option while proposing to "examine the policy and regulatory issues associated with purchase of a commercial reactor or irradiation service." (1995) The cost of this option was estimated by the DOE to be between $2 billion and $4.6 billion in 1996 dollars - a cost to the US government that can be assumed to be less than the cost of subsidizing a Triple Play reactor (O'Leary, 1995). In 1988, the GAO estimated a total construction cost of $2.2 billion for converting the WNP-1 light water reactor at Hanford to produce tritium (GAO, 1988, and Barker, 1988). This would have equated to a cost of about $3 Billion in 1996 dollars.

Conclusion

The lesson to be learned from both the Isaiah and the System 80+ proposals is that a disposition-based conversion project must be both economically and politically feasible or it will not by funded by the US government in the current era of tight budgets. At the present time, these requirements would appear to remove any project involving MOX or reprocessing from consideration as an acceptable vehicle for conversion. As a result, any region desiring to quickly replace lost jobs and decreased economic impact from government facilities that have either closed or sharply reduced their operations would do well to look elsewhere. At a time when it is essential that a region move quickly to replace lost economic base activities, it cannot afford to become wedded to a course of action that is economically unfeasible and politically unwise. To pin a region's hopes on such a proposal invites numerous delays in the conversion process and makes it highly likely that the region will have to restart its search from the beginning after the MOX-based or reprocessing project is finally rejected as a feasible plan.

Note

1   Conversion programs attempt to replace an obsolete or formerly federally funded economic base in a community or region with a new industry or base that does not rely on the old funding sources. In the case of nuclear plant locations, conversion programs have proposed to replace older DOE plants with new, privately-funded nuclear facilities.

References

Barker, Rocky (1988), 'Experts Weigh Four Designs', *The Idaho Falls Post Register*, April 24.

Davis, George (1995), ABB Combustion Engineering, Personal Communication to Brian Costner, Energy Research Associatiates, May.

Dodd, Lauren (1993), *The Isaiah Project*, Battelle Institute, Pacific Northwest Laboratories, WA, October 1.

GAO (1988), Washington, DC.

Honenkamp, John R. (1993), SAIC, *Letter to Dr. Matthew Bunn, National Academy of Science*, November 9.

O'Leary, Hazel (1995), *Briefing on Dual Track Strategy*, Department of Energy, Washington, DC, October 10.

Program Plan For Deployment Of A System 80+ Multi-purpose Nuclear Facility At Savannah River Site (1994), System 80+ Team, Savannah River Site, Aiken, SC, pp. 1-12, 68-75.

Wages, Robert (1993), President, OCAW, *Letter to Elmer Chatak, President, Industrial Union Department*, November 3.

Wald, Matthew L. and Gordon, Michael R. (1994), 'Russia And US Have Different Ideas About Dealing With Surplus Plutonium', *NY Times News Service*, August 19.

Weida, William (1994), *The Political Economy of Nuclear Weapons and Economic Development After the End of the Cold War*, International Congress on Conversion of Nuclear Weapons and Underdevelopment: Effective Projects from Italy, Rome.

Glossary

Accelerator

A device that increases the velocity and energy of charged particles, such as electrons and protons; also referred to as a particle accelerator. In a linear accelerator, particles are accelerated in a straight path.

Actinides

The elements with atomic numbers above 88 (actinium, element 89). The actinide series includes uranium, atomic number 92, and all the man-made transuranic elements. See 'transuranic'.

ALMR

Advanced Liquid Metal Reactor. This is a liquid sodium-cooled reactor.

Atom

A particle of matter indivisible by chemical means. The smallest unit of a chemical element, approximately 1/100,000,000 inch in size, consisting of a nucleus surrounded by electrons.

Atomic nucleus

The central core of an atom, made up of neutrons and protons held together by a strong nuclear force.

Aqueous process

A liquid-based process. Reprocessing that uses liquid chemicals is called aqueous reprocessing.

BNFL

British Nuclear Fuels, Ltd. A British company that reprocesses nuclear materials and produces MOX.

Breeder reactor

A nuclear reactor that produces more fissionable fuel than it consumes.

Burning

The act of using either uranium or plutonium as fuel in a nuclear reactor.

CANDU

The acronym for the Canadian deuterium/uranium reactor.

Closed reactor fuel cycle

A fuel cycle in which spent reactor fuel is reprocessed to recover uranium and plutonium. These materials are then fabricated into new fuel elements that are used to power the reactor.

COGEMA

The state-owned French company in charge of commercial nuclear power, reprocessing of nuclear materials, and MOX production.

Critical

Capable of sustaining a nuclear chain reaction.

Decay heat

The head produced by the decay of radioactive nuclides.

Disposition

The act of disposing of weapon-grade or other fissile materials.

DOE

The US Department of Energy.

Downblending

A procedure to de-enrich highly enriched uranium to make it useable as reactor fuel.

Fast breeder reactor

A fast reactor that produces more fissionable material than it consumes. See 'fast reactor'.

Fast neutrons

Neutrons with energies greater than 100,000 electron volts (considered very high energy).

Fast reactor

A reactor in which the fission chain reaction is sustained primarily by fast neutrons. See 'fast neutrons'.

Fission

The splitting of a nucleus into two approximately equal parts, which are nuclei of other elements, accompanied by the release of a relatively large amount of energy and generally one or more neutrons. Fission can occur spontaneously but usually is caused by nuclear absorption of neutrons.

Fission products

The radioactive fragments (by-products) formed by nuclear fission in a reactor - the 'ash' of nuclear power production. Technetium and iodine radioisotopes are examples of fission products found in spent fuel.

Fuel

Fissionable material used or usable to produce energy in a reactor.

Fuel cycle

The series of steps involved in supplying fuel for nuclear power reactors. It includes mining, refining, enrichment, original fabrication of fuel elements, their use in a reactor, chemical processing to recover fissionable material remaining in spent fuel, enrichment of the fuel material, and fabrication into new fuel elements. Waste disposal is a final step.

Fuel reprocessing

The chemical or metallurgical treatment of spent (used) reactor fuel to recover the unused fissionable material, separating it from radioactive waste. The fuel elements are chopped up and chemically dissolved. Plutonium and uranium and possible other fissionable elements are then separated out for further use.

GWe

Gigawatts (electric). A power rating for nuclear reactors equal to 1000 MWe.

Half-life

The period of time required for the radioactivity of a substance to drop to half its original value; the time that it takes for half of the atoms of a radioactive substance to decay. Measured half-lives vary from millionths of a second to billions of years.

HEU

Highly enriched uranium of the strength normally used in nuclear warheads.

HM

Heavy metal. The general designation of the mix of fission elements present in spent fuel.

INEL

Idaho National Engineering Laboratory. A DOE-controlled nuclear reservation located next to Idaho Falls, Idaho.

Isotope

An isotope of an element is one of two or more forms of the element that differ in their atomic weights (number of neutrons in the nucleus of the element).

kWh

Kilowatt hour.

LEU

Low enriched uranium of the strength normally used for reactor fuel.

Light water

Ordinary water.

LWR

Light water reactor - a reactor using light water for its cooling medium.

Linear accelerator

A long straight tube (or series of tubes) in which charged particles (ordinarily electrons or protons) gain in energy by action of oscillating electromagnetic fields.

Metric ton

One thousand kilograms or about 2200 pounds. The quantities of all nuclear materials are measured in grams, kilograms and metric tons.

Minor actinides

The transuranic elements minus plutonium. Usually this term is used to refer to neptunium, americium, and curium. Some also refer to these as the minor transuranics. Plutonium is the dominant transuranic, but these minor transuranics contribute comparable radioactivity in spent fuel.

MHTGR

Modular High Temperature Gas Cooled Reactor.

MOX

Mixed oxide fuel that contains plutonium and is meant to be used to power reactors.

MWe

Megawatts (electric). A power rating for nuclear reactors equal to 1000 kilowatts.

Neutron

An uncharged particle with a mass slightly greater than that of a proton. The neutron is a strongly interacting particle and a constituent of all atomic nuclei except hydrogen.

Nuclear Reaction

A reaction involving a change in an atomic nucleus, such as fission, fusion, neutron capture, or radioactive decay, as distinct from a chemical reaction, which is limited to changes in the electron structure surrounding the nucleus.

Nuclear reactor

A device in which a fission chain reaction can be initiated, maintained, and controlled. Its essential component is a core containing fissionable fuel. It is sometimes called an atomic furnace; it is the basic machine of nuclear energy.

Nucleus

The central core of an atom, made up of neutrons and protons held together by nuclear force.

Nuclide

Any species of atom that exists for a measurable length of time. The term is used synonymously with isotope. A radionuclide is a radioactive nuclide.

ORNL

Oak Ridge National Laboratory. A DOE controlled nuclear reservation in Oak Ridge, Tennessee.

Open reactor fuel cycle

A fuel cycle in which spent reactor fuel is sent directly to some disposition site for either above ground or below ground storage.

Proton

A particle with a single positive unit of electrical charge and a mass that is approximately 1,840 times that of the electron. It is the nucleus of the hydrogen atom and a constituent of all atomic nuclei.

Pu

The chemical symbol for plutonium. Pu-239 is the isotope of plutonium used to make nuclear weapons.

PUREX process

The plutonium and uranium extraction (PUREX) process is an aqueous process used in several foreign commercial and US defense programs for separating out elements in spent nuclear fuel.

Pyroprocessing

Nonaqueous processing carried out at high temperatures in three steps: electrorefining to separate materials (including actinides) from the radioactive fission products; cathode processing, which further purifies the metal product of the electrorefining; and injection casting to fabricate the reclaimed materials into new forms.

Radioactive

Referring to the spontaneous transformation of one atomic nucleus into a different nucleus or into different energy states of the same nucleus.

Radioactive Decay

The spontaneous transformation of one atom into a different atom or into a different energy state of the same atom. The process results in a decrease, with time, of the original number of radioactive atoms in a sample.

Radioactive Waste

Equipment and materials (from nuclear operations) which are radioactive and for which there is no further use. The waste is generally classified as high-level, low-level, or transuranic, depending on the composition and intensity of the radioactive constituents.

Radioisotope

A radioactive isotope. An unstable isotope of an element that decays spontaneously, emitting radiation. Radioisotopes contained in the spent fuel resulting from the production of nuclear power generally fall into two categories: fission products and transuranic elements (known as transuranics, actinides, or TRU), and activation products produced by neutron absorption in structural materials in the spent fuel.

Recycling

The reuse of fissionable material, after it has been recovered by chemical processing from spent reactor fuel.

Reprocessing

Separating the components of some mix of fissile and spent nuclear materials. Usually this is an aqueous (liquid) process used to separate the components of spent fuel to retrieve plutonium.

S&T technologies

Separation and transmutation technologies.

Spent fuel

Nuclear reactor fuel that has been irradiated (used) to the extent that it can no longer effectively sustain a chain reaction and therefore has been removed from the reactor for disposal. This irradiated fuel contains fission products, uranium, and transuranic isotopes.

SRS

Savannah River Site. A DOE-controlled nuclear reservation in South Carolina.

START

Strategic Arms Reduction Talks.

153

Subcritical

Not capable of sustaining a nuclear chain reaction, but involving some degree of multiplication of neutrons.

SWU

Separative Work Unit. This is a standardized unit of work required in the enrichment of uranium and it is used as a reference figure for pricing enriched uranium.

Target

Material subjected to particle bombardment (as in an accelerator) in order to induce a nuclear reaction.

Thermal neutrons

Low-energy neutrons that have come to thermal equilibrium with the material in which they are moving. Most have energies of less than a few tenths of an electron volt. Current commercial reactors use thermal neutrons.

Thermal cycle plutonium use

The use of plutonium in nuclear reactors.

Transmutation

The transformation (change) of one element into another by a nuclear reaction or series of reactions.

Transuranic

An element above uranium in the Periodic Table of Elements - that is, one that has an atomic number greater than 92. All transuranics are produced artificially (during a man-made nuclear reaction) and are radioactive. They are neptunium, plutonium, americium, curium, berkelium, californium, einsteinium, fermium, mendelevium, nobelium, and lawrencium.

TRUEX

A chemical solvent process under development to extract transuranics from high-level waste.

Waste Separation

The dividing of waste into constituents by type (for example, high-level, low level) and/or by isotope (for example, separating out plutonium and uranium). The waste may be separated by a chemical solvent process such as PUREX or by any of a number of other chemical or physical processes.

Bibliography

'A Great Fixer-Upper' (1996), *The Bulletin of the Atomic Scientists*, Vol. 52, No. 2, March/April, p. 8.

Adinolfi, Roberto. (1992), *The Burning of Uranium Originated by Nuclear Disarmament in Nuclear Power Stations*, Working Papers of the International Symposium on conversion of Nuclear Warheads for Peaceful Purposes, Rome, Italy, June 15,16,17, pp. 56-62.

Albright, David, and Feiveson, Harold A. (1988), 'Plutonium Recycling and the Problem of Nuclear Proliferation', *Annual Review of Energy*, Vol. 13, p. 254.

Albright, David, Franz Berkhout, and William Walker (1993), *World Inventory of Plutonium and Highly Enriched Uranium*, 1992, Oxford University Press, p. 199.

All Things Considered (1996), National Public Radio, December 5.

Ayukawa, Yurika (1996), *Fissile Material Disposition & Civil Use Of Plutonium*, Issue No. 1, yayukawa@igc.apc.org, September 23.

Ayukawa, Yurika (1996), *Fissile Material Disposition & Civil Use Of Plutonium*, Issues No. 2, yayukawa@igc.apc.org, October 3.

Barker, Rocky (1996), 'Experts Weigh Four Designs', *The Idaho Falls Post Register*, April 24, 1988.Berkhout, Franz, *Briefing to Reprocessing Workshop*, October 4.

Berkhout, Franz, Diakov, Anatoli, Feiveson, Harold, Hunt, Helen, Lyman, Edwin, Miller, Marvin and von Hipple, Frank (1993), 'Disposition of Separated Plutonium', *Science and Global Security*, Volume 3, p. 169.

Berkhout, Franz (1993), *Fuel Reprocessing At THORP: Profitability and Public Liabilities*, Center for Energy and Environmental Studies, Princeton University, Princeton, NJ, p. 2.

Berkhout, Franz (1996), 'The Rationale and Economics of Reprocessing', in *Selected Papers Global '95 Concerning Reprocessing*, W.G. Sutcliffe, Ed., UCRL-ID-124105, Lawrence Livermore National Laboratory, California, June 14, pp. 38-39.

Bloomster, C.H., P.L. Hendrickson, M.H. Killinger, and B.J. Jonas (1990), *Options and regulatory issues related to disposition of fissile materials from arms reduction*, PNL-SA-18728, Pacific Northwest Laboratories, US Department of Energy, Washington, DC, pp. 12, 13.

Bradley, Carol (1993), 'Crapo Asks To Fund More Work On Reactor', *The Idaho Statesman*, June 10.

Brown, Paul (1996), 'Production Crisis Hits THORP Nuclear Plant', *Manchester Guardian*, August 27, p. 1.

Buckner, M.R., Radder, J. A., Angelos, J. G., Inhaber, H. (1992), *Excess Plutonium Disposition Using ALWR Technology*, WSRC-RP-92-127B, Westinghouse Savannah River, Aiken, SC, February, p. 10.

'Building A-bombs Requires Less Material Than Had Been Believed, Experts Say' (1994), *New York Times News Service*, August 21.

Bunn, Mathew (1996), Conversation with William Weida , International Conference on Military Conversion and Science: Utilization/Disposal of the Excess Weapon Plutonium: Scientific, Technological and Socio-Economic Aspects, Como, Italy, March 18-20.

Bunn, Matthew (1996), *Getting the Plutonium disposition Job Done: The Concept of a Joint Venture Disposition Enterprise Financed by Additional Sales of Highly Enriched Uranium*, Paper Presented at the International Conference on Military Conversion and Science: Utilization/Disposal of the Excess Weapon Plutonium: Scientific, Technological and Socio-Economic Aspects, Como, Italy, March 18-20, pp. 4, 5.

Campbell, Ronald L., Snider, J. David (1996), *Cost Comparison for Highly Enriched Uranium Disposition Alternatives*, Y/ES-122, Nuclear Materials Disposition Program Office, Y-12 Defense Programs, Lockheed Martin Energy Systems, Inc., April, p. 5.

'Capping Fallout from Russia's Nuclear Legacy' (1995), *The Christian Science Monitor*, November 8.

Carter, Luther J. (1994), 'Lets Use It', *The Bulletin of the Atomic Scientists*, May/June, p. 43.

Charnetski, Joanne and Rauf, Tariq (1994), 'Let Canada Cremate Nuclear Swords', *Defense News*, October 3-9.

Chow, Brian G. and Kenneth A. Solomon (1993), *Limiting the Spread of Weapon-Usable Fissile Materials*, National Defense Research Institute, RAND, Santa Monica, CA, p. 13.

Closing the Circle on the Spitting of the Atom (1996), U. S. Department of Energy, Office of Environmental Management, June, p. 30, 31.

Cochran, Tom, Arkin, William, Norris, Robert, and Hoenig, Mathew (1987), *Nuclear Weapons Databook, Volume II: US Nuclear Warhead Production*, Ballinger, Cambridge, MA, p. 59-64.

Committee on Separations Technology and Transmutation Systems (1996), Board on Radioactive Waste Management, Commission on Geosciences, Environment, and Resources, National Research Council, *Nuclear Wastes: Technologies for Separations and Transmutation*, National Academy Press, Washington, DC, p. ix.

Costner, Brian (1995), Personal communication with George Davis of ABB Combustion Engineering, May.

Cumo, Maurizio (1992), 'General Economic Evaluation of the Conversion of Nuclear Warheads into Electrical Power', *Working Papers of the International Symposium on conversion of Nuclear Warheads for Peaceful Purposes*, Rome, Italy, June 15,16,17, pp. 95-97.

Decressin, Albert, Gambier, Didier J., Lehmann, Jean-Paul, Nietzold, Dieter, E. (1996), *Experience and Activities in the Field of Plutonium Recycling in civilian Nuclear Power Plants in the European Union*, Paper Presented at the International Conference on Military Conversion and Science: Utilization/Disposal of the Excess Weapon Plutonium: Scientific, Technological and Socio-Economic Aspects, Como, Italy, March 18-20, p. 3.

Developing Technology to Reduce Radioactive Waste May Take Decades and Be Costly (1993), GAO/RCED-94-16, United States General Accounting Office, Washington, DC, December, pp. 11, 26.

DeVolpi, Alex (1995), 'Fast finish to plutonium peril', *The Bulletin of the Atomic Scientists*, Vol. 51, No. 5, September/October, pp. 20, 21.

Diakov, Anatoli S. (1996), *Utilization of Already Separated Plutonium in Russia: Consideration of Short- and Long-Term Options*, Paper Presented at the International Conference on Military Conversion and Science: Utilization/Disposal of the Excess Weapon Plutonium: Scientific, Technological and Socio-Economic Aspects, Como, Italy, March 18-20.

Dodd, Lauren (1993), Battelle Institute, *The Isaiah Project*, Pacific Northwest Laboratories, October 1.

Draft Environmental Impact Statement on a Proposed Nuclear Weapons Nonproliferation Policy Concerning Foreign Research Reactor Spent Nuclear Fuel (1995), DOE Office of Environmental Management, US Department Of Energy, Washington, DC, March, pp. 2-16.

Draft Environmental Impact Statement on the Disposition of Highly Enriched Uranium (1995), US Department of Energy, Office of Fissile Materials Disposition, Washington, DC, October.

Duggan, Ruth A., Jaeger, Calvin D., Moore, Lonnie R., and Tolk, Keith M. (1995), 'Non-Proliferation, Safeguards, and Security for the Fissile Materials Disposition Program Immobilization Alternatives', *Final Proceedings: US Department of Energy Plutonium Stabilization and Immobilization Workshop*, December 12-14, p,. 415.

Eldredge, Maureen (1996), *Notes From The September 10, 1996, DOE Briefing*, Military Production Network, in Ayukawa, Yurika, *Fissile Material Disposition & Civil Use Of Plutonium*, Issue No. 1, yayukawa@igc.apc.org, September 23.

Electric Utilities Commentary (1992), 'Are Older Nuclear Plants Still Economic?, Insights from a Lehman Brothers Research Conference', vol. 2, no. 21, May 27, p. i.

'Electricity, The Power Shift Ahead' (1996), *Business Week*, December 2, pp. 78-80.

Energy Statistics Yearbook: 1993 (1995), United Nations, New York, NY.

Federal Energy Subsidies: Direct and Indirect Interventions in Energy Markets (1992), SR/EMEU/92-02, Energy Information Administration, US Department of Energy, Washington, DC, November, p. 7.

Fetter, Steve (1992), 'Control and Disposition of Nuclear Weapons Materials', *Working Papers of the International Symposium on conversion of Nuclear Warheads for Peaceful Purposes*, Rome, Italy, June 15,16,17, pp. 144-148.

Ford, John (1996), *DOE Briefing*, Office of Nuclear Materials and Facility Stabilization, DOE, October 4.

Fuoto, John (1997), Ogden Environmental And Energy Services, personal communication to William J. Weida, jsfuoto@oees.com, January 27.

GAO (1988), Washington, DC.

Garwin, Richard L. (1993), *Critical Question: The Value of Plutonium*, presented at Cornell University, October 1.

Garwin, Richard L. (1992), *Steps Toward the Elimination of Almost All Nuclear Warheads*, Working Papers of the International Symposium on conversion of Nuclear Warheads for Peaceful Purposes, Rome, Italy, June 15,16,17, pp. 17-20.

Garwin, Richard L. (1993), *The Use of Pu for Peaceful Purposes*, August 23, 1993, presented at Cornell University, October 1.

Gingold, J.E., Kupp, R.W., Schaeffer, D., and Klein, R.L. (1991), *The Cost of Reprocessing Irradiated Fuel From Light Water Reactors: An Independent Assessment*, NP-7264, EPRI, Palo Alto, California.

Gray, Leonard, Kan, Tehmau, Shaw, Henry, and Armantrout, Guy (1995), 'Immobilization Needs and Technology Programs', *Final Proceedings: US Department of Energy Plutonium Stabilization and Immobilization Workshop*, December 12-14, pp. 60-61.

Gray, Peter (1994), Personal communication with William J. Weida, June 30.

Grumbly, Thomas, P. (1995), 'Plutonium Stabilization and Immobilization Workshop Objectives', *Final Proceedings: US Department of Energy Plutonium Stabilization and Immobilization Workshop*, December 12-14, p. 18.

Hardung, Heimo (1994), 'Plutonium Naiveté', *Defense News*, April 4-10, pp. 22, 24.

Hartman, H. (1987), *Introductory Mining Engineering*, Wiley, New York, NY.

Heinz, Mark (1996), 'Uranium Prices Rise on Scarcity, Steady Demand', *The Wall Street Journal*, February 26.

Holden, Constance (1993), 'Breaking Up (a Bomb) Is Hard To Do', *Science*, September 24, p. 1673

Honenkamp, John R. (1993), SAIC, Letter to Dr. Matthew Bunn, National Academy of Science, November 9.

Hurt, David (1995), 'Progress on Plutonium Stabilization', *Final Proceedings: US Department of Energy Plutonium Stabilization and Immobilization Workshop*, December 12-14, pp. 36-37.

Ingwerson, Marshall (1995), 'Marketing Nuclear Plants For an Energy-Hungry World', *The Christian Science Monitor*, November 8.

Integral Fast Reactor (1993),Department of Energy, Summer.

Johnson, E.R. (1992), *Alternatives for Disposal of Plutonium from Nuclear Weapons Disarmament Activities*, paper given at the Annual Meeting of the Institute of Nuclear Materials Management, Orlando, Florida, July.

Johnston, J. Bennett (1993), 'IFR For A Safe World', *Letter to the Washington Post*, October 13.

Jouan, Antoine (1995), 'Selecting a Plutonium Vitrification Process', *Final Proceedings: US Department of Energy Plutonium Stabilization and Immobilization Workshop*, December 12-14, pp. 360-363.

Kan, Tehmau, and Sullivan, Kent (1995), 'Glass and Ceramic Immobilization Alternatives and the Use of New Facilities', *Final Proceedings: US Department of Energy Plutonium Stabilization and Immobilization Workshop*, December 12-14, p. 296.

Kass, Jeffrey N. and Erickson, Randy (1995), 'Workshop Perspectives', *Final Proceedings: US Department of Energy Plutonium Stabilization and Immobilization Workshop*, December 12-14, p. 11.

Kerber, Ross (1996), 'Nuclear-Power Plant Shutdown 'Likely' For Facility Run by Northeast Utilities', *The Wall Street Journal*, October 10, p. A6.

'Known Nuclear Tests Worldwide, 1945-1994' (1995), *The Bulletin of the Atomic Scientists*, Vol. 51, No. 3, May/June, p. 70.

Komanoff Energy Associates (1992), *Fiscal Fission: The Economic Failure of Nuclear Power*, 270 Lafayette, Suite 400, New York, NY, December.

Kueppers, Christian and Sailer, Michael (1994), *The MOX Industry or The Civilian Use of Plutonium*, International Physicians for the Prevention of Nuclear War.

Kushnikov, Viktor (1995), 'Radioactive Waste Management and Plutonium Recovery Within the Context of the Development of Nuclear Energy in Russia', *Final Proceedings: US Department of Energy Plutonium Stabilization and Immobilization Workshop*, December 12-14, p. 27.

Leslie, Bret (1994), 'DOE Assesses Dangers from its Plutonium Inventory', *Science for Democratic Action*, Vol. 3, No. 2, IEER, Washington, DC, Spring, pp. 2, 3.

Letter to Tom Clements (1996), Greenpeace International, September 3, 1996 (Contact: 202-319-2506) in Ayukawa, Yurika, *Fissile Material Disposition & Civil Use Of Plutonium*, Issue No. 1, yayukawa@igc.apc.org, September 23.

Lindsay, Richard W. (1994), 'Reactor Solves Problems As It Offers Benefits', *The Idaho Statesman*, Boise, Idaho, June 26.

Lippman, Thomas W. (1993), 'Disputed Nuclear Program Reborn', *The Washington Post*, April 13.

Lippman, Thomas W. (1996), 'US To Burn, Bury Toxic Plutonium From Weapons', *Washington Post*, December 9, p. A01.

Lombardi, Carlo and Mazzola, A. and Vettraino, F. (1996), *Non-Fertile Fuels for Burning Weapons Plutonium in Thermal Fission Reactors*, Paper Presented at the International Conference on Military Conversion and Science: Utilization/Disposal of the Excess Weapon Plutonium: Scientific, Technological and Socio-Economic Aspects, Como, Italy, March 18-20, pp. 5, 6.

Lombardi, Carlo and Mazzola, A. (1994), *Plutonium Burning Via Thermal Fission in Unconventional Matrices*, Paper presented at IAEA Technical Committee Meeting on Unconventional Options for Plutonium Disposition, Obninsk, Russia, November 11.

Lombardi, Carlo (1994), Dipartimento Di Ingegneria Nucleare, Centro Studi Nucleari Enrico Fermi, Letter to William Weida, November 30.

Lombardi, Carlo (1994), *Weapon-grade Plutonium, Annihilation via Thermal Fission in Unconventional Non-fertile Matrices*, International Congress on Conversion of Nuclear Weapons and underdevelopment: Effective Projects from Italy, Rome, 4-5 July.

'Long-Term Program for the Development and Utilization of Nuclear Energy 1994' (1994), *Nuke Info*, No. 41, Tokyo, Japan, June/July.

Lyman, Edwin S. (1995), 'A Perspective on the Proliferation Risks of Plutonium Mines', *Final Proceedings: US Department of Energy Plutonium Stabilization and Immobilization Workshop*, December 12-14, p. 448.

Makhijani, Arjun (1995), 'An Approach to Meeting the Spent Fuel Standard', *Final Proceedings: US Department of Energy Plutonium Stabilization and Immobilization Workshop*, December 12-14, pp. 276-277.

Makhijani, Arjun, and Annie Makhijani (1995), *Fissile Materials In A Glass, Darkly*, IEER Press, Takoma Park, Maryland, p. 10.

Makhijani, Arjun, Schwartz, Steven I. and Norris, Robert S. (1997), 'Retirement and Dismantlement of Nuclear Weapons, and Storage and Disposition of Retired Nuclear Weapons and Surplus Nuclear Weapons Materials', in Schwartz, Steven I, Ed, *Atomic Audit*, Brookings, Washington, DC, pp. 199-201.

Management and Disposition of Excess Weapons Plutonium (Pre Publication Copy) (1994), Committee on International Security and Arms Control, National Academy of Sciences, National Academy Press, Washington, DC, pp. 3, 4.

McWhorter, D. L., Geddes, R.L., Jackson, W. N., and Bugher, W. C. (1995), *Chemical Stabilization of Defense Related and Commercial Spent Fuel at the Savannah River Site*, Document No. NMP-PLS-950239, Westinghouse Savannah River Company, August 16, Sections 3.3, A1.0.

Merrow, E.W., Phillips, K.E., and Meyers, C.W. (1981), *Understanding Cost Growth and Performance Shortfalls in Pioneer Process Plants*, RAND/R-2569j-DOE, RAND Corporation, Santa Monica, California.

National Conference of State Legislatures High-Level Radioactive Waste Newsletter (1996), July.

New Scientist (1995).

Nigon, J. L., and Fournier, W. (COGEMA) (1996), 'MOX Fabrication and MOX Irradiation Experience Feedback from the French Programme', *International Seminar on MOX Fuel: Electricity Generation from Pu Recycling*, United Kingdom, June.

Norris, Robert S. and Arkin, William M. (1995), 'US Nuclear Weapons Stockpile, July 1995', *The Bulletin of the Atomic Scientists*, Vol. 51, No. 4, July-August, pp. 78-79.

Nuclear Energy Agency/Organization for Economic Cooperation and Development (NEA/OECD) (1993), *The Cost of High-Level Waste Disposal: Analysis of Factors Affecting Cost Estimates*, OECD, Paris, p.136.

Nuclear Fuel (1996), September 9, p. 2.

Nuclear Power, Moody's Special Comment (1993), Moody's Investors Service, New York, NY, April, p. 7.

'Nuclear Reactions' (1996), *Washington Post Magazine*, May 5.

Nucleonics Week (1996), April 4.

Nuke Info Tokyo (1995), Citizens' Nuclear Information Center, Japan, No. 50, Nov./DEC.

Numark, Neil J. (1996), *Get SMART: The Case for a Strategic Materials Reduction Treaty, and Its Implications*, Paper Presented at the International Conference on Military Conversion and Science: Utilization/Disposal of the Excess Weapon Plutonium: Scientific, Technological and Socio-Economic Aspects, Como, Italy, March 18-20, p.6.

Office Of Technology Assessment (1977), *Nuclear Proliferation and Safeguards*, (Library of Congress Catalog No. 77-600024), pp. 177, 180, 140.

O'Leary, Hazel (1995), *Briefing on Dual Track Strategy*, Department of Energy, October 10.

Organization for Economic Cooperation and Development (1989), Nuclear Energy Agency, *Plutonium Fuel: An Assessment*, Paris, Table 14.

Organization for Economic Cooperation and Development (1993), Nuclear Energy Agency, *The Economics of the Nuclear Fuel Cycle*, OECD/NEA, Final Revised Draft, Paris.

Panofsky, Wolfgang K.H. (1996), 'No Quick Fix For Plutonium Threat', *The Bulletin of the Atomic Scientists*, Vol. 52, No. 1, January-February, pp. 3, 59.

Passell, Peter (1996), 'Critics say material could fall into terrorists', *The New York Times*, August 28.

Perez, Joseph M., Schumacher, Ray F., and Forsberg, Charles W. (1995), 'Melter Technologies Assessment', *Final Proceedings: US Department of Energy Plutonium Stabilization and Immobilization Workshop*, December 12-14, p. 345.

Peters, W. (1987), *Exploration and Mining Geology*, p. 262.

Peterson, Per F. (1995), 'Long-Term Retrievability and Safeguards for Immobilized Weapons Plutonium in Geological Storage', *Final Proceedings: US Department of Energy Plutonium Stabilization and Immobilization Workshop*, December 12-14, p. 435.

Plutonium--Deadly Gold of the Nuclear Age (1992), International Physicians for the Prevention of Nuclear War and The Institute for Energy and Environmental Research, International Physicians Press, Cambridge, MA, pp. 130-138.

Program Plan for deployment of a System 80+ Multi-purpose Nuclear Facility at Savannah River Site (1994), System 80+ Team, Savannah River Site, Aiken, SC, March 31, pp. 1-12.

Protection and Management of Plutonium (1995), American Nuclear Society Special Report, August, p. 11.

Provost, J.L. (1996), *Plutonium Recycling and Use of MOX Fuel In PWR - French Viewpoint*, Presented at the International Seminar on MOX Fuel: Electricity Generation from Pu Recycling, United Kingdom, June.

Record of Decision, Department Of Energy Programmatic Spent Nuclear Fuel Management and Idaho National Engineering Laboratory Environmental Restoration and Waste Management Programs (1995), DOE Office of Environmental Management and Idaho Operations Office, US Department of Energy, May 30, p. 7.

Rothstein, Linda (1994), 'French Nuclear Power Loses its Punch', *The Bulletin of the Atomic Scientists*, July-August, pp. 8,9.

Rothwell, Geoffrey (1996), *Economic Assumptions for Evaluating Reactor-Related Options for Managing Plutonium*, Paper Presented at the International Conference on Military Conversion and Science: Utilization/Disposal of the Excess Weapon Plutonium: Scientific, Technological and Socio-Economic Aspects, Como, Italy, March 18-20, p. 18-19.

Rougeau, Jean-Pierre (1996), *A Clever Use of Ex-Weapons Material*, Paper Presented at the International Conference on Military Conversion and Science: Utilization/Disposal of the Excess Weapon Plutonium: Scientific, Technological and Socio-Economic Aspects, Como, Italy, March 18-20.

Rudy, Greg (1995), 'Overview of Surplus Weapons Plutonium Disposition', *Final Proceedings: US Department of Energy Plutonium Stabilization and Immobilization Workshop*, December 12-14, p, 22.

Ryabev, L.D., Favorsky, O., Subbotin, V., Kagramanian, V. and Oussanov, V. (1996), *Nuclear Power Strategy of Russia*, Paper Presented at the International Conference on Military Conversion and Science: Utilization/Disposal of the Excess Weapon Plutonium: Scientific, Technological and Socio-Economic Aspects, Como, Italy, March 18-20, p. 2,5.

Sachs, Noah (1996), *Risky Relapse into Reprocessing*, Institute for Energy and Environmental Research, January, p. 20.

Sanger, David E. (1994), Japan, 'Bowing to Pressure Defers Plutonium Projects', *The New York Times*, February 22.

Schneider, Mycle (1996), *Plutonium Fuels*, World Information Service on Energy, Krasnoyarsk, Russia, June, p. 4.

Schulze, Joachim (1992), *Burning of Plutonium in Light Water Reactors (MOX Fuel Elements) Compared To Other Treatment*, Working Papers of the International Symposium on conversion of Nuclear Warheads for Peaceful Purposes, Rome, Italy, June 15,16,17, pp. 65-74.

Schwartz, Steven, ed. (1997), *Atomic Audit*, manuscript to be published by the Brookings Institute.

Seaborg, Glenn T. (1995), Preface to *Protection and Management of Plutonium*, American Nuclear Society Special Report, August.

Siemens Press Release (1995), July 7.

Silver, R. (1996), 'Hydro Puts Off Bruce Retubing As Hope for Pu Mission Fades', *Nucleonics Week*, August 15, p. 8.

Silvestri, Mario (1994), 'Remarks', *International Congress on Conversion of Nuclear Weapons and underdevelopment: Effective Projects from Italy*, Rome, 4-5 July.

Stratton, R. and Bay, H. (1995), *Experience in the Use of MOX Fuels in the Beznau Plants of NOK*, Presented at the International Seminar on MOX Fuel: Electricity Generation from Pu Recycling, United Kingdom, June, 1996.

Synatom (1995), *1995 Report*.

Takagi, Jinzaburo (1996), Citizens' Nuclear Information Center, 1-59-14-302 Higashi-nakano, Nakano-ku, Tokyo 164, Japan, January 10.

Takagi, Jinzaburo (1996), *Japan's Plutonium Program and Its Problems*, Third International Radioecologica Conference on The Fate of Spent Nuclear Fuel and Reality, Krasnoyarsk, Russia, June 22-27, p. 7.

Takagi, Jinzaburo (1995), 'Sodium Leak Hits Achilles' Heel of FBR MONJU', Citizens' Nuclear Information Center, Tokyo, December 9.

Taylor, I.N., Thompson, M.L. and Wadekamper, D.C. (1991), *Fuel Cycle Assessment-1991*, GEFR-00897, General Electric, San Jose, California.

Taylor, Theodore B. (1996), *Solar Disposal of Radioactive Wastes and Plutonium*, Submitted to Science and Global Security, February 15, pp. 24, 25.

Technical Strategy for the Treatment, Packaging, and Disposal of Aluminum-Based Spent Nuclear Fuel (1996), Office of Spent Fuel Management, Department of Energy, June, p. 4.

Technical Summary Report For Surplus Weapons-Usable Plutonium Disposition (1996), DOE/MD-0003, Office of Fissile Materials Disposition, US Department of Energy, Washington, DC, July 17, pp. 4-4 - 4-6, 6-3.

The Oak Ridge Education Project (1992), *A Citizen's Guide to Oak Ridge*, A Project of the Foundation for Global Sustainability, May, p. 7.

Tigner, Brooks (1994), 'Nuclear Bomb Experts: Detonate Stockpile Underground', *Defense News*, September 4, p. 10.

Toevs, J. and Beard, C. (1996), *Gallium in Weapons-Grade Plutonium and MOX Fuel Fabrication*, LA-UR-96-4764, Los Alamos National Laboratory, NM.

Uematsu, Kunihiko (1992), 'The Technological and Economic Aspects of Plutonium Utilization in Fission Reactors', *Working Papers of the International Symposium on conversion of Nuclear Warheads for Peaceful Purposes*, Rome, Italy, June 15,16,17, pp. 109-115.

Uranium Institute (1996), Editor/contact: Jack Ashton, at *1996 NucNet*, nucnet@atagbe.ch, August 26.

US Congress, Office of Technology Assessment (OTA) (1993), *Technologies Underlying Weapons of Mass Destruction*, OTA-BP-ISC-115, US Government Printing Office, Washington, DC, December, p. 156.

US Department of Energy (1996), *DOE Realizes Post-Cold War Dividend While Meeting 'START' Targets*, April 18.

US Department of Energy (1994), *Openness Press Conference Fact Sheets*, June, 27, pp. 172-173.

van Vliet, J., Haas, D., Vanderborck, V., Lippens, M., and Vandeberg, C (Belgonucleaire) (1996),*MIMAS MOX Fuel Fabrication & Irradiation Performance*, International Seminar on MOX FUEl: Electricity Generation from Pu Recycling, United Kingdom, June.

Vargas, Dale (1993), 'Scientists tout new nuke reactor as safe', *Sacramento Bee*, May 17.

Vendreyes, Georges (1992), 'Plutonium Burning In Fast Reactor and As MOX Fuel', *Working Papers of the International Symposium on conversion of Nuclear Warheads for Peaceful Purposes*, Rome, Italy, June 15,16,17, pp. 61-64.

Vendreyes, Georges (1994), *Energy for Mankind and the Use of Fissile Material from Disarmament in Nuclear Reactors*, International Congress on Conversion of Nuclear Weapons and underdevelopment: Effective Projects from Italy, Rome, 4-5 July.

von Hippel, Frank, and Marvin Miller, Harold Feiveson, Anatoli Diakov, Frans Berkhout (1993), 'Eliminating Nuclear Warheads', *Scientific American*, August, pp. 48.

von Hipple, Frank (1992), 'Control and Disposition of Nuclear Weapons Materials', *Working Papers of the International Symposium on conversion of Nuclear Warheads for Peaceful Purposes*, Rome, Italy, June 15,16,17, pp. 119-128.

von Hipple, Frank (1995), 'Fissile Material Security In The Post-Cold-War World', *Physics Today*, June, p. 32.

von Hipple, Frank (1995), *Reprocessing of Spent Power-Reactor Fuel: Why We Can Wait, Why We Should Wait*, International Conference on Evaluation of Emerging Nuclear Fuel Cycle Systems, Versailles, France, September 12, pp. 2-3.

Wages, Robert (1993), President, OCAW, Letter to Elmer Chatak, President, Industrial Union Department, November 3.

Wald, Matthew L. (1996), 'Agency To Pursue 2 Plans To Shrink Plutonium Supply', *The New York Times*, December 10.

164

Wald, Matthew L. (1994), and Michael R. Gordon, 'Russia And US Have Different Ideas About Dealing With Surplus Plutonium', *NY Times News Service*, August 19.

Weida, William (1994), *The Political Economy of Nuclear Weapons and Economic Development After the End of the Cold War*, International Congress on Conversion of Nuclear Weapons and Underdevelopment: Effective Projects from Italy, Rome, 4-5 July.

Printed and bound by CPI Group (UK) Ltd, Croydon, CR0 4YY

23/10/2024

01778241-0005